Functional Foods
Processing and Packaging

NEW INDIA PUBLISHING AGENCY
New Delhi-110 034

About the Editors

Prof. (Dr.) Tanweer Alam is currently working as the Director (I.C) of Indian Institute of Packaging (IIP), Mumbai. Before joining IIP, he has worked as Assistant Professor at Banaras Hindu University, Varanasi, SHUATS, Allahabad and Jiwaji University, Gwalior. He has been working for last two decades in different academic institutions and also involved in teaching of graduate, postgraduate and doctoral students. His Doctoral research work at the prestigious National Dairy Research Institute, Karnal was based on Modified Atmosphere Packaging of food products. He had been guest faculty at University of Copenhagen, Denmark under faculty exchange programme funded by European Union. He performed as jury for Asia Star Award and World Star Award. He has organized several conferences/workshops of national/international stature for promoting packaging industry. He has published several research and technical papers in high impact referred journals, apart from delivering invited lectures both in India and abroad. He has also contributed more than 20 book chapters and also edited dozens of proceedings and technical souvenir and authored edited five books. He is Associate Editor of Asian Journal of Dairy Science. Editor in cheif of Journal of Packaging Technology and Research and also associated with Editorial Board of Processed Food Industry. He is Life member of different scientific and academic societies.

Prof. Saket Kushwaha, currently the Vice Chancellor of Rajiv Gandhi Central University, Arunachal Pradesh, is specialized in resource management and sustainable agriculture development. After higher studies from Banaras Hindu University (BHU), joined Abubakar Tafawa Balewa University (ATBU), Bauchi, Nigeria in 1993 and taught various courses in Agriculture Economics, Agribusiness, Agriculture Extension and Management at undergraduate, post graduate and Ph.D. level. Prof. Kushwaha rose to the rank of Professor at ATBU in the year 1999 and in 2006 joined BHU, India as professor in Agriculture Economics and became the Vice Chancellor of Lalit Narayan Mithila University, Bihar for one term 2014-2017. Prof Kushwaha has more than 120 publications in national and international journals of repute to his credit, supervised 24 Ph.D. students and authored 21 books / book chapters. Prof. Kushwaha is life member of 10 Professional Bodies and sits in the panel of several editorial boards. He has worked extensively in the field of Zero Emission Research Initiative (ZERI) propagating the mission of sustainable development under the aegis of "Waste is Wealth" concept. He is also the recipient of 19 national and international awards which includes award from Sulabh International

Gold Medal in 2016 for sanitation management. Prof Kushwaha has 27 years of teaching, research, extension and community service experience with 23 years in administration as Head of the Department, Coordinator, Prof-in-Charge (Director), Dean of Faculty and Vice Chancellor. He has successfully concluded over 10 national and international projects majorly funded by World Banks. He has coordinated USAID project on Cowpea Research Support Programme (CRSP) in Nigeria for 10 years from 1996 to 2006. He is working with NGOs and mega agriculture farms for Green Farm Planning.

Dr. Arun Kumar is presently working as Associate Prof. and Head of Department of Dairy and Food Chemistry, College of Dairy and Food Technology, Maharana Pratap University of Agriculture and Technology, Udaipur, Rajasthan since 2007. Dr. Kumar has obtained his M.Sc. and Ph.D degree in Dairy Chemistry from National Dairy Research Institute, Karnal (Haryana), Asia's premier institute in Dairying.

Dr. Arun Kumar served as a Senior Analyst in the state of the art NABL accredited Food Testing Laboratory, FICCI's Research and Analysis Centre (FRAC), New Delhi sponsored by FICCI and Ministry of Food Processing and Industries, New Delhi for four and half years. He has handled Ministry of Food Processing Industries, New Delhi, India sponsored R&D project entitled "Characterization of Colostrum Bio-active Components from Different Species and their application in the Formulation of Novel Dairy Products of amount 87.26 lacs. He has also acted as Course Director of ICAR, New Delhi sponsored Winter school entitled "Significance of Bioactive Ingredients and Supplements in Health Foods". He has successfully contributed in research findings of six sponsored R & D projects. He has guided 55 MSc. and Ph.D students. He has published 32 research articles in national and international journals of repute. He has 12 awards to his credit in recognition of his academic and research excellence by prestigious institutes. He has participated in various national and international conferences, delivered radio talks, TV Shows as an expert on Dairy and Food Quality Aspects. He is a life member of several professional societies like Dairy Technology Society of India, NDRI, Karnal, Indian Dairy Association (IDA), New Delhi, Indian Journal of Animal Science, New Delhi, Rajasthan Chapter of Asian Agri History Foundation, Udaipur, AFST etc and a regular reviewer of Springer Journals.

Prof. (Dr.) Sahar Masud is presently working as Professor (Fisheries) at SKUAST, Jammu. Presently She is Head of Division of Fisheries Resource Management, Integrated Fish farming and Aquaculture.

She has guided Post Graduate and Ph.D students and having Publications in the field Fisheries Resource Management, Integrated Fish farming, organic farming and Aquaculture. Authored books on Prospects and Potential of Fish Farming in Jammu and Kashmir, How to start Fish farming, Vision document andpolicy paperfor development of Fisheries and Aquaculture in Jammu and Kashmir.

Functional Foods Processing and Packaging

Tanweer Alam
Director
Indian Institute of Packaging (IIP)
Mumbai, Maharashtra

Saket Kushwaha
Vice Chancellor
Rajiv Gandhi Central University
Arunachal Pradesh

Arun Kumar
Associate Prof. and Head
Department of Dairy and Food Chemistry
College of Dairy and Food Technology
Maharana Pratap University of Agriculture and Technology
Udaipur, Rajasthan

Sahar Masud
Professor and Head
Division of Fisheries Resource Management
Sher-e-Kashmir University of Agricultural Sciences and Technology, Jammu
Jammu, Jammu and Kashmir

NEW INDIA PUBLISHING AGENCY
New Delhi-110 034

NEW INDIA PUBLISHING AGENCY
101, Vikas Surya Plaza, CU Block, LSC Market
Pitam Pura, New Delhi – 110 034, India
Email: info@nipabooks.com
Web: www.nipabooks.com

For customer assistance, please contact
Phone: + 91-11-27 34 17 17 Fax: + 91-11- 27 34 16 16
E-Mail: feedbacks@nipabooks.com

ISBN: 978-93-89571-76-9

Composed and Designed by NIPA.

कृषि वैज्ञानिक चयन मंडल
कृषि अनुसंधान और शिक्षा विभाग
कृषि एवं किसान कल्याण मंत्रालय, भारत सरकार
कृषि अनुसंधान भवन I, पूसा, नई दिल्ली 110 012
AGRICULTURAL SCIENTISTS RECRUITMENT BOARD
DEPARTMENT OF AGRICULTURAL RESEARCH & EDUCATION
Ministry of Agriculture & Farmers' Welfare, Govt. of India
KRISHI ANUSANDHAN BHAVAN I, PUSA, NEW DELHI 110 012

प्रो. (डॉ.) ए. के. श्रीवास्तव
अध्यक्ष
Prof. (Dr.) A K Srivastava
Chairman

Dated: March 30, 2022

Foreword

The fast paced life in changing times has given rise to stress and related diseases. Such a situation affects the sufferer both physically and mentally and in turn proves detrimental to the nation's health. Further, the changing food habits with more dependence on processed food has resulted in a very significant change in the food constituents of consumers. Therefore, it is necessary to understand the effect that different types of food constituents have on our body physiology. It is rightly understood that the overall well-being of a person is dependent on the type of food that one consumes.

Research in food constituents and its pharmacological effect on our health is not something new, though it has come in focus lately, particularly with the rise of life-style diseases. It is an acknowledged fact that all societies and communities have their own indigenous knowledge of food derived from the study and observation of nature and her gifts bestowed upon us. Studies in functional foods and Nutraceuticals have taken this process steps further and today we can be aware of the exact effect of most of the food that we consume. The interest in nutraceuticals and functional foods has risen immensely in the recent past due to their potential nutritional, and therapeutic effects.

The present book "Functional Foods: Processing and Packaging" by Prof. Tanweer Alam, Prof. Saket Kushwaha, Dr. Arun Kumar and Dr. Sahar Masud stands in good stead at this crucial juncture of time when the world is bracing to protect themself against all sorts of life-style dependent diseases. The book will be very useful for food scientists, researchers and technologists engaged directly in the field of nutraceuticals, functional foods and related areas. To readers it will provide a wide range of information on various aspects of functional foods for maintaining the good health. The contributing authors to this volume are experts in their own fields. It gives me immense pleasure to congratulate the editors and all the contributing authors for their efforts in bringing out this timely and valuable book.

Srivastava

(Dr. A.K. Srivastava)

कृषि अनुसंधान भवन-I, पूसा, नई दिल्ली 110 012 | Krishi Anusandhan Bhavan I, Pusa, New Delhi 110 012
Tel: 011-25842536 *Email:* aksrivastava.asrb@gmail.com; memberasrb.as@gmail.com

Preface

Nutraceutical and functional foods has gained immense popularity in the whole world to combat with many diseases. In recent days, people from India and other developing countries spends more time and effort to maintain health through exploring various herbal medicines and natural health care products. The World Health Organization on March 11, 2020, has declared the novel coronavirus (COVID-19) outbreak a global pandemic. Since then, the Severe Acute Respiratory Syndrome Coronavirus 2(SARS-CoV-2) has caused, and is continuing to cause, global disruption, a significant loss of life and extensive human suffering. Studies reveal that nutraceutical and functional foods which include vitamin C, vitamin D, omega-3-polyunsaturated fatty acids, probiotics, and zinc can largely be attributed towards preventive healthcare with the possibility of inhibiting the replication of SARS-CoV-2 and its prognosis. With the ongoing situation, a large part of the population now understands the role of these immunity boosting supplements and this has led to a significant change in consumer behaviour. Moreover, it has been seen from ancient time that herbal medicines play a great role in preventing diseases and offer better health to human being. India is the origin of "Ayurveda"means "The Science of Life" and is often called the "Mother of All Healing". Herbal medicine is the basis of "Ayurveda medicine" which comes into limelight over 5000 years ago in India. Many nutraceuticals and functional foods are worth to mention here which are used to develop herbal medicines. Moreover, special micronutrients (Phytochemicals) are discovered which are used to cure many diseases like cardiovascular disease (CVD), cancer, diabetes, dementia etc. Nature is enriched with such ingredients which are progressively discovered by many research groups. Several research works are already pursued to explore the relationships among various food components, their physiological functionality and health benefits. Functional foods contain natural extracts, concentrates, or natural ingredients majorly which can reduce current healthcare costs by health improvement and disease prevention. The past decade has evidenced consumer driven trend of the functional food consumption. Future prospect of this trend is excellent and as a result, scientific information on all aspects of functional food ingredients is important for the advancement of this emerging sector. The second wave of COVID-19 has also paved the way for the nutraceutical sector to build a strong presence in the domestic as well as global market.With an objective to provide better understanding of use, benefits, significance and impact on human health for various nutraceutical foods, natural bioactive compounds and to disseminate the recent developments in such a rapidly expanding field, this book entitled "Functional Foods – Processing and Packaging" under the publication of the New India Publishing Agency, New Delhi has been compiled and edited. Seventeen chapters in this book not only cover various aspects of functional foods and bioactive compounds from diverse natural sources and its impact, but also discuss the sources and applications of natural antioxidants, probiotics, prebiotics and synbiotics.The contributing authors are experts in their respective fields, and I am grateful to each and every one of them for their thoughtfulness in contributions to this book. This book will be of interest to a wide spectrum of professionals from food scientists and technologists, nutritionists, biochemists, and engineers to entrepreneurs worldwide. It will also serve as a unique reference for food scientists for the R&D departments of food companies that are working with functional foods and ingredients. Additionally, it will serve as a source of basic information for college and university students majoring in food science and technology, food processing, and engineering. Readers will obtain sound scientific knowledge about various aspects of nutraceuticals and functional foods or food ingredients, fermented functional food, various natural bioactive compounds and antioxidants.

Editors

Contents

1

Functional Foods Concept Trends and its Packaging Aspects

Donal Bhattacharjee

R&D Division, Indian Institute of Packaging, Mumbai, Maharashtra

Introduction

The health and wellness continue to drive growth in the global food industry as the consumers are constantly looking for the foods, beverages and other dietary solutions that will help them more aggressively to meet their own unique nutrition and personal health goals. Worldwide countries are going through a nutritional transition and are now affected by double burden of problems like nutritional and food security. The common alarming problems are under nutrition and specific nutrient deficiencies simultaneously coupled with imbalanced diets and chronic diseases. The problem of chronic diseases, non-communicable diseases (cardiovascular diseases, respiratory diseases, diabetes, obesity, cancer etc.) and malnutrition is more serious in the developing countries. The lack of access to health services, sufficient income, healthy and nutritious foods characterize majority of poor people in developing countries. So, food is not only the solution but also viewed as a product to enhance health and wellbeing, and it encourages the producers by supplying new goods that meet these needs. From the past decade, there has been growing consumer awareness regarding health, nutrition and the food safety which are often triggered by specific food incidents, especially in the developed countries and high-income groups in such countries. In the food industry there have been two broad trends which are hazards in the food chain and nutritional profile of the food. There has been recognition that hazards and contaminants can enter the food chain and can cause sickness, death and diseases, and so governments, often to deal with consumer and industry concerns, establish regulations and standards on food hazards including standards on pesticides, antibiotics, microbiological hazards and other chemical and physical contaminants. On the other hand, there is also a growing awareness of the health-enhancing properties for prevention and treatment of health concerns beyond the basic nutritional component of many foods.

Over the years, the consumption pattern of consumers is emerging. Studies shows that the consumers demand for food that helps prevents disease, boost mental health, and improve the quality of life. The global demand for functional foods is growing annually. The food components and physiological mechanisms can be manipulated to reap the greatest benefits from their properties. Presently, the focus is moving beyond disease prevention to optimal health, optimal performance and maximum wellbeing. In addition to staying healthy, consumers seek to enhance functions such as mental or physical performance, perceived energy level, appearance, and even mood. The foods which we consume contain various dietary components with an array of health benefits that offer an excellent opportunity to improve public health and well-being. The nutritional benefits of foods have been explored since time immemorial. So, considering the background, a consumer is only likely to consider switching conventional with functional food if the latter is perceived as healthier in comparison. The awareness of functional food is growing, and its demand is increasing even in developing countries. The propelling forces for the demand of functional food includes increasing health conscious, growing health cost and awareness about the value of functional food. Hence, in this chapter, we analyzed the concept and perception about the functional foods.

What are Functional Foods?

All foods are functional to some extent because all foods provide taste, aroma and nutritive value. Foods are now being examined intensively for added physiologic benefits, which may reduce chronic disease risk. The consumption of plant-derived foods like fruits, vegetables, and whole grains has a wide range of health benefits and the phytochemicals present in the them have been associated with health benefits are glucosinolates, sulfur containing compounds of the Alliaceae, terpenoids (carotenoids, monoterpenes, and phytosterols), and various groups of polyphenols (anthocyanins, flavones, isoflavones, stilbenoids, ellagic acid, etc.). This wide range of products cannot be truly classified as "food" and can be designated as functional foods. The concept of functional foods is often considered to have emerged in Japan in the late 1980s. However, functional foods actually have a quite long history. In China, Japan and other Asian countries, many types of foods have traditionally been associated with specific health benefits (Kobayashi et al., 2017; Butnariu and Sarac, 2019). Functional foods have been considered as foods made up of nutrients that confer physiological or medical benefit to the consumers. In fact, any healthy food confirmed to have a health promoting or disease preventing property beyond the basic functions of supplying nutrition is now regarded as a functional food (Martirosyan and Singh, 2015). Nonetheless, it is important to note that there is no universally accepted regulatory definition of the term 'functional foods' to date, despite their wide usage in marketing, and they represent one of the fastest growing segments of the food industry in most nations of the world. Since, functional food emerged in nutrition research, other terms such as: "nutraceuticals, designer foods, pharmafoods, medifoods, vitafoods, dietary supplements, and fortified foods" have appeared. This complicates nutrition vocabulary and confuses consumers. The Functional Food Center (FFC) at Dallas, USA has been a pioneer in the functional food industry, combining cutting-edge expertise in the biomedical sciences with practical business experience in order to aid in further research, development, and commercialization of functional food innovations in both domestic and international markets. The FFC connects a global network of professionals (scientists, functional food experts, and food industry representatives) to an environment conducive to innovative research collaboration. Functional foods are part of normal dietary food intake for optimized nutrition.The Functional Food Center defines "functional foods" as ***"Natural or processed foods that contain biologically-active compounds; which, in defined, effective, non-toxic amounts, provide a clinically proven and documented health benefit utilizing specific biomarkers, for improving general health, for the prevention, management, or treatment of chronic disease or its symptoms."***

In the current definition, it is observed that functional foods can be natural or processed. It means that foods in their pure unchanged or changed forms can be functional. A natural unchanged food has not undergone human interference. A changed food has undergone some chemical modification, whether it is the addition, removal, alteration, or increased bioavailability of a chemical compound within the food (Roberfroid, 2000). For example, a natural food may be a pure orange or avocado, while a processed food may be folate-fortified cereal or milk infused with Vitamin C. The FFC defines food as components of a normal diet for optimized nutrition. This includes conventional foods and not pills or capsules. The functional foods contain known or unknown biologically-active (bioactive) compounds. Biologically-active compounds or secondary metabolites are molecules in food, usually in small amounts, that act synergistically to benefit health. As stated above, the active components of functional foods may be "an essential macronutrient, if it has specific physiological effects, or an essential micronutrient, if its intake is over and above the daily recommendations" (Anju et al., 2014). A micronutrient (e.g. vitamins or minerals) is required and must be ingested in certain trace amounts. A macronutrient (e.g. carbohydrates, proteins, and fats) must be consumed in large amounts as they make up most of human diet. Specifically, bioactive compounds like flavonoids, capsaicinoids, lignin, trepenoids, carotenoids, chlorophylls, vitamins, stilbene, phenolic acids, fibers, sterols, lipids, fatty acid, polysaccharides, and some plant-derived proteins and peptides may exert antioxidant, cardio-protective, and chemo-preventive effects, which slow disease development or progression (Kris-Etherton et al., 2002; Abugri and Melissa, 2014). The humans may need to obtain certain nutrients by ingesting bioactive compounds because humans cannot synthesize these nutrients naturally, such as chlorophyll from lettuce.

The functional foods containing bioactive compounds must be consumed in certain effective non-toxic amounts. This fact is paramount to functional foods working therapeutically and not toxically. For example, Vitamin C is a known bioactive compound that should be consumed at 90 mg. in order to maintain normal health. A food containing more than 200-300 mg. of Vitamin C per day may act therapeutically in terms of reducing one's risk of contracting a cold, thereby acting as a functional food. However, there comes a point at which consuming Vitamin C becomes toxic, which is at approximately 2000 mg. or more daily (Pacier and Danik, 2015).The functional foods are meant to protect against chronic illness, and therefore, must contain the correct dosage of the bioactive compound in question, neither too little nor too much. Such functional foods must provide a clinically proven and documented health benefit. Demonstrating success in clinical studies or tests involving humans are imperative to functional food acceptance among health experts, policy makers, and the public. In order to demonstrate causal relationships between functional foods and the prevention, management, or treatment of chronic disease, extensive pre-clinical and clinical trials must be performed. The current definition of functional food highlights the importance of "bioactive compounds" within functional foods. Bioactive compounds are considered the backbone of functional food effectiveness. Functional food scientists frequently use biomarkers in their research to determine the rate or effectiveness of a biological process in its natural state as well as after functional food administration. When a bioactive compound is proposed to exhibit certain benefits, changes in biomarker activity confirms or denies these benefits. Biomarkers can manifest in the form of protein, blood sugar, cholesterol, triglyceride levels, or more subtly, the levels of hormone stimulating particular tissue. The biomarkers can indicate the mechanism by which bioactive compounds prevent or treat illness.

Therefore, in addition to providing new options for improving health and well-being, the functional foods sector offers potential for new economic opportunities. As Japan is the birthplace of the term 'functional food' and the country has been at the forefront of the development of functional foods since the early 1980s when systematic and large-scale research programmes were launched and funded by the Japanese government on systematic analysis and development of food functions. As a result the concept of foods for specified health use (FOSHU) was established in 1991. These foods, which are intended to be used to improve people's health and for which specific health effects (claims) are allowed to be displayed, are included as one of the categories of foods described in the Nutrition Improvement Law as foods for special dietary use.

Functional food is a relatively new concept to Indian consumers. Food Safety and Standard Act (FSSA) (2006) is the single reference point in relation to regulation of functional foods in India. However, Food Safety and Standards Authority of India (FSSAI) (2006) definition is relevant in Indian context. Broadly "Functional food" may be defined as a food which influences specific functions in the body that may provide added health benefits or remedy from some diseased condition following the addition/concentration of a beneficial ingredient, or removal/substitution of an ineffective or harmful ingredient. The functional food market in India is growing at a rapid pace as health as a value has percolated in the Indian society undergoing dramatic demographic shifts as people's awareness of influence of diet on health has been increasing. The driving forces like rising awareness about health and fitness, ageing population, changing lifestyle are fostering this growth(Sharma and Garg, 2013). The most popular FFPs are energy or sports drinks, probiotic dairy products, heart health spreads and ready-to-eat cereals (Weststrate et al., 2002).

Why Functional Foods?

The initiative to develop functional foods has arisen from the growing interest in the relationship between diet, specific food ingredients and health. Healthy eating can make a key contribution to health and well-being, but due to present lifestyle the busy consumers may not have the time to access their optimal diet. Functional foods can be solution as it provides health enhancing ingredients in a convenient form. The increased affluence and urbanization is also linked to a lifestyle where our daily routine requires less physical activity and there is greater access to foods with higher energy densities. As a result, energy intake and expenditure are not in balance leading as us becoming overweight. Further, the prevalence of chronic non-communicable diseases such as cardiovascular diseases (CVD), high blood pressure and type-2 diabetes has increased, leading to concerns about escalating costs of healthcare and reduced quality of life. Modern lifestyles and longer life are also linked to various mental health problems such as depression, poor concentration and loss of memory. Therefore, challenge is to develop effective synergies between science and food product development for the benefit of the consumer.

Types of Functional Foods

The functional foods are often supplemented with bioactive substances and are expected to decrease the risk of diet-related disorders and diseases. There are several classifications of functional foods. The most frequently used is the classification into (a) Natural- foods or product that naturally contain bioactive substances, (b) Processed-processed foods that have been supplemented with bioactive substances during production, and (c) Manipulated-changed foods with an increased content of bioactive components obtained via modification such as genetic engineering and animal feeding (Martirosyan and Singh, 2015). The examples of bioactive components are dietary fiber; omega-3 polyunsaturated fatty acids (PUFA); phytoestrogens; polyphenols; lycopene; carotenoids such as alpha- and beta-carotene; lutein and zeaxanthin and plant sterols and stanols (Fiedor and Burda, 2014; Weaver, 2014; Perez-Gregorio and Simal-Gandara, 2017; Serino and Salazar, 2019).

Dietary Fiber

Dietary fiber is a group of food components resistant to human digestive enzymes and found mostly in foods of plant origin. Foods rich in dietary fiber include cereal products, seeds of leguminous plants, fruits, and vegetables with various amounts and quality of fiber. The main fiber fractions are hemicelluloses, pectins, cellulose, lignin, and β-glucans. Dietary fiber contributes to lowering the concentration of glucose and cholesterol in the blood by binding bile acids and cholesterol (Evans et al., 2015; McRae, 2017).

Omega-3 Polyunsaturated Fatty Acids

The foods enriched in omega-3 PUFAs are of two major classes: omega-3 and omega-6 fatty acids (Fas) (de Goede et al., 2010). Omega-3 FAs supplementation benefits the cardiovascular system in both patients with CVD and healthy individuals. A significant source of omega-3 FA are fish, such as herring, bluefish, mackerel, sardines, salmon, tuna, trout, and oil from anchovy. Major plant sources are flaxseeds, chia seeds, canola and soybean oil, walnuts, algae, sea buckthorn (*Hippophaer hamnoides*), camelina (*Camelina sativa*), and avocados (Jędrusek-Golińska et al., 2020). PUFAs are indispensable for the brain, heart, and other organs development and function, especially in infancy and older age (Nestel et al., 2015). High dosages of omega-3 FA (>4 g/day) from marine food sources lowered the concentration of triglycerides in the blood by 9% to 26% in individuals with normolipidemia to borderline hyperlipidemia and small anti-arrhythmic and hypotensive effect (Leslie et al., 2015). PUFAs have also been described as having neuronal protective functions and providing benefits in the prevention of dementia (Phillips et al., 2012; Eriksdotter et al., 2015; Yanai, 2017). Functional food products enriched with omega-3 Fats are oils, margarine, bread, bread spreads, pasta and spaghetti, chocolate, milk drinks, and fruit juices (Kolanowski and Laufenberg, 2006). A specific limitation in the production of products with fish oil is oxidative and sensory changes occurring during storage. Fish oil is also added to the farm animals' forage, which causes the increase of the long-chain omega-3 FAs content in eggs, milk, or meat fat, allowing them to be classified as functional foods (Bourre, 2005).

Phytoestrogenic Compounds

Phytoestrogens are plant compounds synthesized from phenylpropanoids and simple phenols with structural similarity to estrogens and affinity to estrogen receptors. Major groups of phytoestrogens are isoflavones, coumestans, and lignans (Rietjens et al., 2017). Major sources of isoflavones are soybeans and its products containing from approximately 3.3 to 12.2 mg/g, and red clover (He and Chen, 2013). Major isoflavones such as genistein, daidzein, glycitein, biochanin A and formononetin are converted in the digestive tract to compounds with some hormonal activity. Isoflavones display either estrogenic or antiestrogenic affects depending on the serum estradiol concentration. Isoflavones showed estrogenic effects when the endogenous estradiol concentration is low; otherwise, they might have an anti-estrogenic effect (Ding et al., 2016). Isoflavones can also improve hyperglycemia, glucose tolerance, and circulating insulin concentration (Babu et al., 2013; Kushwaha, 2014).

Natural Antioxidants

In humans, oxidative stress has a significant role in the development of cancer, CVD, arthritis, autoimmune, and neurodegenerative diseases. For example, the processes of lipids and lipoproteins oxidation, significantly contribute to atherosclerotic changes in humans (Tomkin and Owens, 2012). A vital role in preventing these changes is performed by vitamins A, C, and E, and by flavonoids (Gilbert, 2013; Rizvi et al., 2014; Perez-Gregorio and Simal-Gandara, 2017). Thus, enriching the diet with antioxidant substances that "capture" free radicals might play an important role in delaying cardiovascular and neurological system diseases, some cancers, cataracts, and other non-communicable diseases prevalent in older adults (Pellegrino, 2016; Liguori et al., 2018). The main dietary sources of flavonoids and vitamins A, C, and E are vegetables (for example, onions, tomatoes, peppers, and broccoli), fruits (for example, citrus fruits, apples, berries, black currants, and grapes), some cereals and spices (for example, turmeric), red wine, tea (especially green), coffee, and cocoa. Particularly rich in flavonoids are red wine, pomegranate, cranberry and grapefruit juice, and dark bitter chocolate (>70% cocoa) (Pandey and Rizvi, 2009). Flavonoids can affect oxidative stress by preserving the hemodynamic and biochemical functions of the cell. They indirectly chelate copper and iron ions, which prevents the formation of hydroxyl radicals, interrupts the free radical reactions cascade in enzymatic and non-enzymatic lipid peroxidation, and protects low molecular weight antioxidants (for example, ascorbate in the cytosol and α-tocopherol in biological membranes) against oxidation. Thus, flavonoids could potentially decrease the risk for CVD, neurological diseases, some cancers, and diabetes type 2 (Perez-Gregorio and Simal-Gandara, 2017). The most common functional foods enriched in flavonoids are drinks, fruit juices, vegetable juices, margarine, and breakfast cereals.

Lutein and Zeaxanthin

Lutein and zeaxanthin are dietary carotenoids that are important factors in eye health as well as preventing age-related cataracts and macular degeneration (AMD) (Bone et al., 2011; Abdel-Aal et al., 2013). Lutein and zeaxanthin are synthesized and present in the green leafy, yellow, and orange vegetables, corn, some wheat species, and in smaller amounts in some fruits (Perry et al., 2009). The primary dietary sources of lutein are spinach, kale, parsley, romaine lettuce, and egg yolk. In the human body, lutein can react with free radicals and protects the low density lipoprotein (LDL) against oxidative processes, associated with the risk of atherosclerosis and ischemic heart disease (Ribaya-Mercado and Blumberg, 2004). The carotenoids, especially lutein, are essential in the prophylaxis of age-related eyesight deterioration such as AMD and cataracts (Johnson, 2014). It is estimated that one in 10 persons aged 65 to 75 years suffers from secondary vision disorders (Sin et al., 2013). Lutein from eggs is three times better bioavailable by humans, compared to lutein from other sources (Abdel-Aal et al., 2013). Eggs and eggs products, and corn products such as cornmeal, corn flour, cornflakes and canned corn are good sources of lutein and zeaxanthin.

Probiotics and Prebiotics

Probiotics are defined as single or mixed cultures of live microorganisms, which, when ingested in adequate amounts, can alter intestinal microbiota and provide some health benefits to the host (Tiihonen et al., 2010). Probiotic products contain live microorganisms in quantity above the dose recommended to obtain health benefits (5×10^9 CFU/day for 5 days) (Gupta and Garg, 2009). Prebiotics are defined as food components, which can selectively stimulate the growth and/or activity of intestinal bacteria strains capable of fermenting them, for example, non-digested starch, oligosaccharides, and non-starch polysaccharides (Upadhyay and Moudgal, 2012). The most common probiotics including strains of *Lactobacillus* and/or *Bifidobacterium* and less frequently *Escherichia*, *Enterococcus*, or *Bacillus* are yogurts, kefir, buttermilk, acidophilus milk, fermented cabbage, pickles, tempeh, miso, and kombucha (Zawistowska-Rojek et al., 2016). The non-digestible oligosaccharides such as fructooligosaccharides (inulin), galactooligosaccharides, xylooligosaccharides, cyclodextrins, and lactulose are recognized as prebiotics. The familiar sources of oligosaccharides are chicory root, Jerusalem artichoke, raw garlic, onions, and whole wheat foods (Al-Sherajiae et al., 2013; Kuntz et al., 2013). Probiotic use has been associated with improved immune function and probiotics have been used in the treatment of gastrointestinal disorders such as constipation, diarrhea, and managing of Crohn's Disease, irritable bowel syndrome (IBS), and diverticular disease. Probiotics improved microbiota composition (Pandey et al., 2015; Clements and Carding, 2018; Stavropoulou and Bezirtzoglou, 2019). The products with probiotics include fermented dairy products (yogurt, kefir, buttermilk, and others), cheeses, and kombucha. Probiotic bacteria are present in some other food products such as ice cream, frozen desserts, fermented fruit, vegetable juices (Zareba and Ziarno, 2011), and fermented meats (Kołożyn-Krajewska and Dolatowski, 2009). The food market also offers products supplemented with prebiotics, such as fermented milk drinks, yogurts, fruit drinks; bread spread fats, meat products, mayonnaise, margarine, and pasta (Liutkevičius et al., 2016).

Sterols and Stanols

Functional foods with cholesterol-lowering potential have attracted much interest as possible alternative therapies for lowering total plasma cholesterol in older individuals and, thus, in the prevention of CVD in older individuals. The consumption of foods containing plant sterols which have the structure of steroid alcohols and include phytosterols (e.g. β-sitosterol, campesterol and stigmasterol) and phytostanols (Scholle et al., 2009; Ogbe et al., 2015) is one option, another being plant sterols as supplements. Functional products enriched with plant sterols include margarine spreads, cream cheese, yogurt and yoghurt drinks, maturing cheeses, milk drinks, mayonnaise, meat products, soy or rice-based drinks, salad and herb sauces, chocolate, and rye bread.

Table 1: Examples of Functional Food Components

S.N.	Component/ Class		Benefits	Source
1	Dietary Fiber	Insoluble fiber	Supports maintenance of digestive health; may reduce the risk of some types of cancer	Wheat bran, corn bran, fruit skins
		Soluble fiber	May reduce risk of coronary heart disease and some types of cancer	Psyllium seed husk, peas, beans, apples, citrus fruits
		Beta glucan	May reduce risk of coronary heart disease	Oat bran, oatmeal, oat flour, barley, rye
		Whole grains	May reduce risk of coronary heart disease and some types of cancers; supports maintenance of healthy blood glucose levels	Cereal grains, whole wheat bread, oatmeal, brown rice
2	Fatty Acids	Monounsaturated fatty acids (MUFAs)	May reduce risk of coronary heart disease	Tree nuts, olive oil, canola oil
		Polyunsaturated fatty acids (PUFAs) Omega-3 fatty acids- Alpha linolenic acid	Supports maintenance of heart and eye health; supports maintenance of mental function	Walnuts, flaxseeds, flaxseed oil
		PUFAs – Omega-3 fatty acids- docosahexaenoic acid (DHA)/ Eicosapentaenoic acid (EPA)	May reduce risk of coronary heart disease; supports maintenance of eye health and mental function	Salmon, tuna, marine and other fish oils
		Conjugated linoleic acid (CLA)	Supports maintenance of desirable body composition and immune health	Beef and lamb; some cheese

S.N.	Component/ Class		Benefits	Source
3	Vitamins	Vitamin A	Supports maintenance of eye, immune and bone health; contributes to cell integrity	Organ meats, milk, eggs, carrots, sweet potato, spinach
		Thiamine (Vitamin B1)	Supports maintenance of mental function; helps regulate metabolism	Lentils, peas, brown or enriched white rice, pistachios and certain fortified breakfast cereals
		Riboflavin (Vitamin B2)	Supports cell growth; helps regulate metabolism	Lean meats, eggs, green leafy vegetables, dairy products and certain fortified breakfast cereals
		Niacin (Vitamin B3)	Supports cell growth; helps regulate metabolism	Dairy products, poultry, fish, nuts, eggs and certain fortified breakfast cereals
		Pantothenic acid (Vitamin B5)	Helps regulate metabolism and hormone synthesis	Sweet potato, organ meats, lobster, soybeans, lentils and certain fortified breakfast cereals
		Pyridoxine (Vitamin B6)	Supports maintenance of immune health; helps regulate metabolism	Beans, nuts, legumes, fish, meat, whole grains and certain fortified breakfast cereals
		Folate or folic acid (Vitamin B9)	May reduce birth defects like brain or spinal cord; supports maintenance of immune health	Beans, legumes, citrus fruits, green leafy vegetables and fortified breads, cereals, rice
		Cobalamin (Vitamin B12)	Supports maintenance of mental function; helps regulate metabolism and supports blood cell formation	Eggs, meat, poultry, milk and certain fortified breakfast cereals
		Biotin (Vitamin H)	Helps regulate metabolism and hormone synthesis	Liver, salmon, dairy, eggs, oysters and certain fortified breakfast cereals
		Ascorbic Acid (Vitamin C)	Neutralizes free radicals which may damage cells; supports maintenance of bone and immune health	Guava, sweet red/green pepper, kiwi, citrus fruit, strawberries, fortified foods and beverages
		Calciferol(Vitamin D)	May reduce the risk of osteoporosis; helps regulate calcium and phosphorus; supports immune health; helps support cell growth	Sunlight, fish, fortified foods such as yogurts or cereals, and beverages, including milk and juices
		Tocopherol (Vitamin E)	Neutralizes free radicals, which may damage cells; supports maintenance of immune and heart health	Sunflower seeds, almonds, hazelnuts, turnip greens, fortified foods and beverages

S.N.	Component/ Class		Benefits	Source
4	Minerals	Calcium	May reduce the risk of osteoporosis	Sardines, spinach, yogurt, low-fat dairy products, fortified foods and beverages
		Magnesium	Supports maintenance of normal muscle and nerve function, immune health and bone health	Spinach, pumpkin seeds, whole grain breads and cereals, halibut, almonds, brazil nuts, beans
		Potassium	May reduce the risk of high blood pressure and stroke, in combination with a low sodium diet	Potatoes, low-fat dairy products, whole grain breads and cereals, citrus juices, beans, banana, leafy greens
		Selenium	Neutralizes free radicals which may damage cells; supports maintenance of immune and prostate health	Fish, red meat, whole grains, garlic, liver, eggs
5	Flavonoids	Anthocyanins – Cyanidin, Pelargonidin, Delphinidin, Malvidin	Bolster cellular antioxidant defences; supports maintenance of healthy brain function	Berries, cherries, red grapes
		Flavanols – Catechins, Epicatechins, Epigallocatechin	Supports maintenance of heart health	Tea, cocoa, chocolate, apples, grapes
		Procyanidins and Proanthocyanidins	Supports maintenance of urinary tract health and heart health	Cranberries, cocoa, apples, strawberries, grapes, red wine, peanuts, cinnamon, tea, chocolate
		Flavanones – Hesperetin, Naringenin	Neutralizes free radicals which may damage cells; bolster cellular antioxidant defences	Citrus fruits
		Flavonols – Quercetin, Kaempferol, Isorhamnetin, Myricetin	Neutralizes free radicals which may damage cells; boost cellular antioxidant defences	Onions, apples, tea, broccoli
6	Isothiocyanates	Sulforaphane	May enhance detoxification of undesirable compounds; bolsters cellular antioxidant defences	Cauliflower, broccoli, broccoli sprouts, cabbage, kale, horseradish

S.N.	Component/ Class		Benefits	Source
7	Phenolic Acids	Caffeic acid, Ferulic acid	Strengthens cellular antioxidant defences; supports maintenance of eye and heart health	Apples, pears, citrus fruits, some vegetables, whole grains, coffee
8	Plant Stanols/ Sterols	Free Stanols/Sterols	May reduce risk of coronary heart disease	Corn, soy, wheat, fortified foods and beverages
		Stanol/Sterol esters	May reduce risk of coronary heart disease	Stanol ester dietary supplements, fortified foods and beverages, including table spreads
9	Polyols	Sugar alcohols – Xylitol, Sorbitol, Mannitol, Lactitol	May reduce risk of dental caries	Some chewing gums another food applications
10	Probiotics	Yeast, *Lactobacilli, Bifidobacteria* and other specific strains of beneficial bacteria	Supports maintenance of digestive and immune health; benefits are strain specific	Certain yogurts and other cultured dairy and non-dairy applications
11	Prebiotics	Inulin, Fructooligosaccharides (FOS), Polydextrose	Supports maintenance of digestive health; supports calcium absorption	Whole grains, onions, some fruits, garlic, honey, leeks, banana, fortified foods and beverages
12	Phytoestrogens	Isoflavones – Daidzein, Genistein	Supports maintenance of bone and immune health, and healthy brain function; for women, supports menopausal health	Soybeans and soy-based foods
		Lignans	Support maintenance of heart and immune health	Flax seeds, rye, some vegetables, seeds and nuts, lentils, triticale, broccoli, cauliflower, carrot
13	Soy Protein	Soy Protein	May reduce risk of coronary heart disease	Soybeans and soy-based foods like milk, yogurt, cheese and tofu
14	Sulfides/ Thiols	Diallylsulfide, Allyl methyl trisulfide	May enhance detoxification of undesirable compounds; supports maintenance of heart, immune and digestive health	Garlic, onions, leeks, shallots
		Dithiolthiones	May enhance detoxification of undesirable compounds; supports maintenance of healthy immune function	Cruciferous vegetables

(*Source*: International Food Information Council Foundation, 2011)

Genetically Engineered Functional Foods

Genetic engineering has been increasingly applied to secure nutritional enhancement of the crops as the need of nutritional food security became apparent as a major health risk in the world (Shukla and Mattoo, 2010). While there has been some hesitation with regard to the acceptability and adoption of biotechnology products in certain developing countries. For example, the introduction of several genes helped develop high-b-carotene rice, now commercially known as Golden Rice. The use of carotene desaturase from *Erwinia uredovora* resulted in a 23-fold increase in total carotenoids in rice (Paine et al., 2005). Other phytonutrients whose content has been enhanced by genetic engineering include α-tocopherol, vitamin C corn etc. (Cho et al., 2005; Kim et al., 2005; Tavva et al., 2007; Yusuf and Sarin, 2007; Crowell et al., 2008; Naqvi et al., 2009). Furthermore, it provides insights into the potential for application of biotechnology in developing improved quality and functional foods for human nutrition and health, rather than simply for use in agricultural technology for improved yields and pest control. Some examples of genetically modified crops with functional components are shown in Table 2.

Table 2: Examples of Genetically Modified Crops with Functional Components

Sl. No.	Functional Compound/ Substance	Crop	Transgene	Benefit
1	Provitamin A	Rice	Phytoene synthase (daffodil); Phytoene desaturase (*Erwinia*); Lycopene cyclase (daffodil)	Antioxidant, Vitamin A supplement
2	Vitamin E	Canola	Tocopherol methyl transferase (Arabidopsis)	Antioxidant
3	Flavonoids	Tomato	Chalcone isomerase (Petunia)	Antioxidant
4	Fructans (indigestible polysaccharide)	Sugarbeet	1-sucrose:sucrose fructosyltransferase (*Helianthus tuberosus*)	Low calorie
5	Iron	Rice	Ferritin (*Phaseolus*), Phytase (mutant, Aspergillus)	Iron supplement

(Ramesh et al., 2012)

Packaging of Functional Foods

The effect of packaging on the variation in food composition during storage depends on the technology used and the storage conditions. As a rule of thumb, functional foods packaging and preservation practices are based on the conditions applied to products of similar origin and format. This approach can give satisfactory results for several products but should not be taken as a universal rule regarding all functional foods. In some cases, the usual technologies and packaging materials could be detrimental to the shelf-life of certain functional ingredients. However, with the major focus on functional ingredients, label design and marketing aspects, the product packing often takes a back seat. The packaging can even end up as little more than an after-thought in development and marketing process of new functional food products, which can be a costly oversight. Recent advances in packaging open up an array of possibilities to ultimately deliver better functionality, quality and taste to the consumer. The application of new packaging technologies can increase shelf life and improve the safety of fresh products such as meats and fish. Other applications include theft protection for premium/high value products such as meat, fish, cheese and dietary supplements. With increased emphasis on reducing the carbon footprint there are also opportunities to produce greener functional foods by making the packaging biodegradable, reducing the amount of material used to make the packaging and reducing transport costs.

Packaging Requirements for Functional Foods

The major chunk of functional foods is of plant origin which comprises of fruits and vegetables. The processing and preservation technologies have an extensive effect on the vitamins, minerals and other phytochemicals like polyphenols, flavanoids, prostaglandins, and sulfur compounds. Phenolic compounds are closely associated with the nutritional quality of foods especially due to their antioxidant capability. These antioxidant compounds easily oxidize in the presence of oxygen. Phenolic compounds are present in cereals, legumes, oilseeds, fruits, and beverages of plant origin like tea, cocoa, beer, wine, fruit juices, etc. Processed plant products with medium-to-high water activity are most susceptible to browning and phenolic degradation (Maltini et al., 2003; Elizaveta, 2012; Venir and Maltini, 2013). So, the conditioning and packaging of these products require the control of oxygen level. However, minimal processing is less detrimental than high temperature treatments and is often a way to preserve the bioactive constituents in functional foods. Modified atmosphere packaging (MAP) is a minimal process well known for its capacity to enhance the shelf-life of fresh fruits and vegetables (Goswami and Mangaraj, 2011; Arvanitoyannis, 2012). The shelf-life improvement obtained using MAP is usually associated with the preservation of normal appearance and firmness. MAP slows down or matches the metabolism of packaged fresh products and microflora to avoid fermentation and thus minimizes microbial growth (Nicola and Fontana, 2014). In case of the probiotics, the viability of microorganisms may be affected by several factors, packaging conditions being one of them. The integration of probiotics into foods involved the use of dried cultures. Dried probiotic cultures are affected by processing and storage conditions such as drying, freezing, moisture, light and oxygen (Fu and Chen, 2011; Peighambardoust et al., 2011; Gong et al., 2014; Poddar et al., 2014). Numerous functional food products can be classified as intermediate moisture foods (IMF) such as cakes, syrups, toaster pastries, fruit rolls, and soft bars. The intermediate moisture levels in such foods are generally related to moisture content of 15%–40% and water activity of 0.65–0.85. Despite the relative shelf stability at ambient temperature, the rates of some deterioration processes are very high at an intermediate moisture level (Rao et al., 2012). Lipid oxidation is also a problem in IMF, provoking rancidity and unacceptable sensory defects (Kader, 2009). The extracted natural oils and fats contain molecules with unsaturated compounds and natural antioxidants like tocopherol. These molecules are easily oxidized when exposed to oxygen and light. The mechanisms of degradation of the bioactive constituents in packaged functional foods are mainly related to UV sensitivity, water activity, microbial load, and oxidation processes. Based on these considerations, the packaging technologies should be selected for their potential to control light, humidity level, temperature and atmosphere conditions. Table 3 briefly summarized some factors to be taken into account in the choice of appropriate packaging conditions for different types of functional foods.

Table 3: Packaging Requirements for Functional Foods

Sl. No.	Functional Food Products	Deterioration Factor	Deterioration Effect on Functional Compound	Packaging Requirements
1	Fresh fruits and vegetables	Moisture, temperature	Vitamin degradation	Controlled and modified atmosphere
2	Processed fruits and vegetables	Oxygen, temperature, microorganism	Loss of nutritional quality, Degradation of antioxidants and polyphenols	High barrier
3	Fermented milks	Oxygen, light, temperature	Vitamin oxidation, decrease in survival of probiotic bacteria, oxidation of vitamins	Controlled oxygen and light barrier
4	Dried probiotics	Moisture	Decrease in the active load	Moisture barrier
5	Protein-based products	Light	Modification of proteins and amino acids	Light barrier
6	Oils and fats	Oxygen, light	Lipid oxidation, rancidity,	Oxygen and light barrier
7	Sprouts	Microorganisms	Flavor problems, Pathogen development	Moisture control

(Louise, 2016)

Innovative Packaging Solutions for Functional Foods

The simple solution to extend the shelf-life of functional foods includes gas flushing, barrier, controlled atmosphere, modified atmosphere, and vacuum packaging. However, new advanced packaging approaches could provide additional tools to increase the shelf-life of bioactive functional food constituents.

Active packaging

Active packaging is a class of technologies in which dynamic physical and/or chemical interactions between the content and the containers induce modifications of the internal conditions resulting in shelf-life improvement. Active packaging can positively influence food preservation by integrating more sophisticated systems of oxygen scavenging, release of synthetic or bioactive compounds, moisture and atmosphere control, and antimicrobial surface properties (Imran et al., 2010). The active packaging includes gas absorbers for oxygen, carbon dioxide, ethylene, and water vapor scavenging. Polymer films can also be incorporated with UV-absorbent agent (Goswami and Mangaraj, 2011). Antioxidant release was one of the first active packaging systems to be implemented in the industry. The phenolic compounds are added to the food contact layer which act as oxygen scavengers and chelators, the most popular additives being butylated hydroxyanisole, butylated hydroxytoluene, and tertiary butyl hydroquinone (Decker et al., 2010). Cereal, nut, and oil-packaging materials often contain such additives to increase their storage stability. For functional foods, complex chains of chemical interactions involving vitamins, polyphenols, and enzymes are critical. The release of antioxidants naturally involved in the biochemical processes would present an interesting strategy. New active packaging systems are using natural molecules as antioxidants. These antioxidants have been integrated in polymer films in the form of pure molecules such as α-tocopherol, carvacol, and thymol or in the form of natural extracts (Colin-Chavez et al., 2013). Functional natural antioxidants are themselves considered as interesting additives in active food packaging systems (Castro-Lopez Mdel et al., 2014). The advantages of using natural compounds in packaging materials could be particularly beneficial to the preservation of functional foods in enhancing stability, bioavailability, and sensory quality of these products.

Nanotechnology

Nanotechnology is a newly emerging technique, which involves the characterization, fabrication, and manipulation of structures, devices or materials that have at least one dimension having 1-100 nm in length (Duncan,

2011). This technology deals with nanomaterials and nanosystems commonly smaller than 100 nanometers. Nanomaterials are defined as materials with any external dimension on the nanoscale, and are clustered into three classes, namely nanoparticles, nanofibers and nanoplates. The interest of using nanomaterials is related to their unique characteristics, only possible due to their nanosize. Nanotechnology can bring several advantages such as improved mechanical and barrier properties, thermal stability, and better optical properties. Apart from that it has great potential to improve packaging materials' main properties as well as adding new functionalities, such as active (e.g., antimicrobial) and intelligent (controlled release) features. Therefore, nanotechnology can be used to overcome some of the main problems of conventional packaging like low gas barrier and controlled delivery of bioactive compounds (Cerqueira et al., 2018).This section is not intended to cover all aspects of nanotechnologies related to food packaging and safety. The intention here is to present some examples of nano-based packaging technologies that could be of interest in preserving quality and shelf-life of functional foods. A variety of nanomaterials such as silver nanoparticle, titanium nitride nanoparticle, and nano-titanium dioxide, nano-zince oxide, and nanoclay are introduced as functional additives to food packaging (Tager, 2014). Edible nano-coatings formulated from functional and bioactive food-grade components are particularly interesting in the field of functional foods. Edible coatings applied directly onto the food are already currently used to provide barrier protection and improve the shelf-life and quality of a large variety of foods, from fresh to processed products (Weiss et al., 2006). Nanotechnologies can also be associated to smart packaging under the form of time-temperature indicators (TTI), gas detectors, and nano-sensors (Mihindukulasuriya and Lim, 2014; Ranjan et al., 2014). TTI systems are intended to be used as abuse indicators. They are usually based on a color change when a certain condition is broken. The property of colloidal gold nanoparticles to irreversibly agglomerate under freezing has been used by Timestrip® to develop a visual accidental freezing indicator for chilled foods (Kagan et al., 2007). The most promising and implementable nano-packaging innovations are related to sensor incorporation for the detection of food deterioration and monitoring storage conditions, increasing barrier properties and strength of materials, oxygen scavenging, and prevention of growth of pathogens.

Future of Functional Food Packaging

The expanding reliance on e-commerce, the need for products to stand out on crowded supermarket shelves, continued demands for sustainability, and advancements in protective properties are all potential game changers for brands looking to succeed in the functional food space. The e-commerce supply chain poses challenges to minimizing expense, maximizing speed of delivery, and enhancing protection of the product. This means that brands must focus on weight, size, and durability. Packaging must not add weight. The importance of reducing bulk and weight in packaging underscores the need to utilize highly durable materials that can provide adequate protection against the elements, withstand a jostling alongside other products, and arrive at the consumer's door in the high-quality condition they expect. Damaged packaging is a turnoff for consumers and deters them from making repeat purchases. The product and brand credibility is important to a consumers' purchase decision, and it is essential for functional food companies to invest in the right packaging to communicate this attribute. The sustainability is inherently connected to many other packaging attributes, including functional design and cost efficiency. Over time, a heightened sense of environmental responsibility has driven consumers to become more conscious of sustainable packaging than ever. They are wary of products that are packaged with too much material and appear wasteful. So, these are the major challenges for the market of the functional foods.

Conclusion

Presently people are increasingly conscious about cognitive health, weight management, and overall health and wellbeing. Such awareness makes way for the functional foods which can be an important part of healthy lifestyle, balanced diet and regular physical activity. But the success of functional foods depends on several factors like their effectiveness, safety, and quality. Apart from that, the bioavailability of these functional food components and the levels required in humans are critical factors necessary to optimize health benefits. It is interesting that the functional foods are sold at premium prices, thus functional food product industry is highly lucrative than any conventional food sector. This excess profit is the main attraction for the food giants and other key players towards the functional food industry. Moreover, the shelf-life of functional foods is critically dependent on their storage and packaging conditions that are applied throughout the entire distribution chain. These conditions have to be precisely adapted to the specific needs of each and every functional food. Unfortunately, research on packaging conditions effects on functional foods is lacking. To optimize the preservation of these sensitive products, there is an urgent need for appropriate data. Nevertheless, the establishment of specific labeling regulations and standards to functional foods will hopefully stimulate further research in this area. It is recommended that nanopackaging should be designed in such a way to release antimicrobials, antioxidants, flavors enzymes and nutraceuticals to extend the shelf life of the food products. However, before going to develop the whole market for functional food products, companies needed to give focus on their efforts to make aware of consumers and communicating the health claims clearly to make them trust on such food products. Future research is needed to shed light on the above raised issues and more information is required for consumers to make them better choices of their diets that contain functional food components.

References

Abdel-Aal, E.S.M., Akhtar, H., Zaheer, K. and Ali, R. (2013). Dietary sources of lutein and zeaxanthin carotenoids and their role in eye health. Nutrients, 5:1169–1185. https://doi.org/10.3390/nu5041169

Abugri, D.A. and Melissa, J. (2014). Efficacy of Bioactive Compounds and Determine Intake Level Necessary to Achieve Desired Effect. Print.

Al-Sherajiae, S.H., Amin, I., Mohd, M.Y., Shuhaimi, M., Rokiah, M.Y. and Fouad, A.H. (2013). Prebiotics as functional foods: A review. Journal of Functional Foods, 5(4):1542-1553. https://doi.org/org/10.1016/j.jff.2013.08.009

Arvanitoyannis, I.S. (2012). Principles of MAP and definitions of MAP, CA and AP. In Modified Atmosphere and Active Packaging Technologies, Ed. I.S. Arvanitoyannis. CRC Press, Boca Raton, FL.

Babu, P.V., Liu, D. and Gilbert, E.R. (2013). Recent advances in understanding the anti-diabetic actions of dietary flavonoids. The Journal of Nutritional Biochemistry, 24(11): 1777-1789. https://doi.org/10.1016/j.jnutbio.2013.06.003

Bone, R.A., Landrum, J.T., Beatty, S. and Nolan, J. (2011). Targeting AMD with a critical carotenoid. Review of Ophthalmology, 7:91-94.

Bourre, J.M. (2005). Where to find omega-3 fatty acids and how feeding animals with diet enriched in omega-3 fatty acids to increase nutritional value of derived products for human: What is actually useful? The Journal of Nutrition Health and Aging, 9(4):232-242.

Butnariu, M. and Sarac, I. (2019). Functional Food. International Journal of Nutrition,3(3):7-16.

Castro-Lopez Mdel, M., Lopez-Vilarino, J.M., Gonzalez-Rodriguez, M.V. (2014). Analytical determination of flavonoids aimed to analysis of natural samples and active packaging applications. Food Chemistry, 150(1):119-127.

Cerqueira, M.A., António, A., Vicente, Lorenzo, and Pastrana, M. (2018). Nanomaterials for Food Packaging: Materials, Processing Technologies, and Safety Issues Micro and Nano Technologies, Elsevier Inc. Pages 1-11.

Cho, E.A., Lee, C.A., Kim, Y.S., Baek, S.H., De Los Reyes, B. and Yun, S.J. (2005). Expression of α-tocopherolmethyltransferase transgene improves tocopherol composition in lettuce (*Lactuca sativa* L.). Molecules Cells, 19:16-22.

Clements, S.J. and Carding, S.R. (2018). Diet, the intestinal microbiota, and immune health in aging. Critical Reviews in Food Science and Nutrition, 4(58):651-661. https://doi.org/10.1080/10408398.2016.1211086

Colin-Chavez, C., Soto-Valdez, H., Peralta, E., Lizardi-Mendoza, J., Balandran- Quintana, R. (2013). Diffusion of natural astaxanthin from polyethylene active packaging films into a fatty food simulant. Food Research International, 54(1):873-880.

Crowell, E.F., McGrath, J.M. and Douches, D.S. (2008). Accumulation of vitamin E in potato (*Solanum tuberosum*) tubers. Transgenic Res, 17:205-217.

de Goede, J., Geleijnse, J.M., Kromhout, D. and Verschuren, W. M. (2010). Marine (n-3) fatty acids, fish consumption, and the 10-year risk of fatal and nonfatal coronary heart disease in a large population of Dutch adults with low fish intake. Journal of Nutrition, 140(5):1023-1028. https://doi.org/10.3945/jn.109.119271

Decker, E.A., Chen, B., Panya, A. and Elias, R.J. (2010). Understanding antioxidant mechanisms in preventing oxidation in foods. In: Oxidation in Foods and Beverages and Antioxidant Applications: Understanding Mechanisms of Oxidation and Antioxidant Activity, Vol. 1, eds. Decker, E., Elias, R., and McClements, D.J. Woodhead Publishing, Cambridge, UK.

Dhiman, A., Walia, V. and Nanda, A. (2014). Introduction to the Functional Foods. In: Introduction to Functional Food Science: Textbook. 2nd Ed. Richardson, TX: Functional Food Center.

Ding, M., Pan, A., Manson, J. E., Willett, W. C., Malik, V., Rosner, B. and Sun, Q. (2016). Consumption of soy foods and isoflavones and risk of type 2 diabetes: A pooled analysis of three US cohorts. European Journal of Clinical Nutrition, 70(12):1381-1387. https://doi.org/10.1038/ejcn.2016.117

Duncan, T.V. (2011). Applications of nanotechnology in food packaging and food safety: barrier materials, antimicrobials and sensors. J Colloid Interface Sci, 363:1-24.

Elizaveta, S. (2012). Water activity concept and its role in food preservation. Meridian Ingineresc, 4:40-48.

Eriksdotter, M., Vedin, I., Falahati, F., Freund-Levi, Y., Hjorth, E., Faxen-Irving, G. and Wahlund, L.O. (2015). Plasma fatty acid profiles in relation to cognition and gender in Alzheimer's disease patients during oral omega-3 fatty acid supplementation: The OmegAD Study. Journal of Alzheimer's disease, 48(3):805-812. https://doi.org/10.3233/JAD-150102

Evans, C.E.L., Greenwood, D.C., Threapleton, D.E., Cleghorn, C.L., Nykjaer, C., Woodhead, C. E. and Burley, V.J. (2015). Effects of dietary fibre type on blood pressure: A systematic review and meta-analysis of randomized controlled trials of healthy individuals. Journal of Hypertension, 33(5):897-911. https://doi.org/10.1097/HJH.0000000000000515

Fiedor, J. and Burda, K. (2014). Potential role of carotenoids as antioxidants in human health and disease. Nutrients, 6(2):466-488. https://doi.org/10.3390/nu6020466

Fu, N. and Chen, X.D. (2011). Towards a maximal cell survival in convective thermal drying processes. Food Research International, 44(5):1127-1149.

Gilbert, C. (2013). What is vitamin A and why do we need it? Community Eye Health Journal, 26(84):65-67.

Gong, P., Zhang, L. and Han, X. (2014). Injury mechanisms of lactic acid bacteria starter cultures during spray drying: A review. Drying Technology, 32(7):793-800.

Goswami, T.K. and Mangaraj, S. (2011). Advances in polymeric materials for modified atmosphere packaging (MAP). In: Multifunctional and Nanoreinforced Polymers forFood Packaging. Woodhead Publishing Limited, Cambridge, UK.

Gupta, V. and Garg, R. (2009). Probiotic. Indian Journal of Medical Microbiology, 27(3), 202-209. https://doi.org/10.4103/0255-0857.53201

He, F.J. and Chen, J.Q. (2013). Consumption of soybean, soy foods, soy isoflavones and breast cancer incidence: Differences between Chinese women and women in Western countries and possible mechanisms. Food Science and Human Wellness, 2(3–4):146-161. https://doi.org/10.1016/j.fshw.2013.08.002

Imran, M., Revol-Junelles, A.M. and Martyn, A. (2010). Active food packaging evolution: Transformation from micro- to nanotechnology. Critical Reviews in FoodScience and Nutrition, 50(9):799-821.

Jędrusek-Golińska, A., Górecka, D., Buchowski, M., Wieczorowska-Tobis, K., Gramza-Michałowska, A. and Szymandera-Buszka, K. (2020). Recent progress in the use of functional foods for older adults: A narrative review. Compr Rev Food Sci Food Saf. 1-23.

Johnson, E.J. (2014). Role of lutein and zeaxanthin in visual and cognitive function throughout the lifespan. Nutrition Reviews, 72(9):605-612. https://doi.org/10.1111/nure.12133

Kader, A.A. (2009). Effects on nutritional quality. In: Modified and Controlled Atmospheres for the Storage, Transportation, and Packaging of Horticultural Commodities, Ed. E.M. Yahia. CRC Press, Boca Raton, FL.

Kagan, M., Mirsky, V. and Wolfbeis, O. (2007). Indicateur de froideur irreversible. Google Patents EP 2035796 A2.

Kim, Y.J., Seo, H.Y., Park, T.I., Baek, S.H., Shin, W.C., Kim, J.G., Choi, Y.E. and Yun, S.J. (2005). Enhanced biosynthesis of a-tocopherol in transgenic soybean by introducing g-TMT gene. J Plant Biotechnol, 7:1-7.

Kobayashi, E., Sato, Y., Umegaki, K. and Chiba, T. (2017). Analysis of Safety Alerts Associated with Dietary Supplements from Japan and Overseas. Food Hygiene and Safety Science (Shokuhin Eiseigaku Zasshi), 59(2):93-99.

Kolanowski, W. and Laufenberg, G. (2006). Enrichment of food products with polyunsaturated fatty acids by fish oil addition. European Food Research and Technology, 222(3–4):472-477. https://doi.org/10.1007/s00217-005-0089-8

Kołożyn-Krajewska, D. and Dolatowski, Z.J. (2009). Probiotics in fermented meat products. ActaScientiarumPolonorum, TechnologiaAlimentaria, 8(2):61-74.

Kris-Etherton, P.M., Hecker, K.D., Bonanome, A., Coval, S.M., Binkoski, A.E., Hilpert, K.F., Griel, A.E. and Etherton, T, D. (2002). Bioactive compounds in foods: their role in the prevention of cardiovascular disease and cancer. American Journal of Medicine, 113:71-88.

Kuntz, M.G.F., Fiates, G.M.R. and TeIxeira, E. (2013). Characteristics of prebiotic food products containing inulin. British Food Journal, 115(2):235-251. https://doi.org/ 10.1108/ 00070701311302212

Kushwaha, M. (2014). Role of soy isoflavones in human health. International Journal of Chemical and Biological Sciences, 1:50-55.

Leslie, M.A., Cohen, D.J.A., Liddle, D.M., Robinson, L.E. and Ma, D.W.L. (2015). A review of the effect of omega-3 polyunsaturated fatty acids on blood triacylglycerol levels in normolipidemic and borderline hyperlipidemic individuals. Lipids in Health and Disease, 14:53. https://doi.org/10.1186/s12944-015-0049-7

Liguori, I., Russo, G., Curcio, F., Bulli, G., Aran, L., Della- Morte, D. and Abete, P. (2018). Oxidative stress, aging, and diseases. Clinical Interventions in Aging, 13:757-772. https://doi.org/10.2147/CIA.S158513

Liutkevičius, A., Speičien˙e, V., Vaičiulyt˙e-Funk, L., Mieželien˙e, A., Alenčikien˙e, G., Abaravičius, A. and Kaminskas, A. (2016). An inulin-enriched soy drink and its lowering effect on oxidized low density lipoproteins in healthy volunteers. Polish Journal of Food andNutrition Sciences, 66(1):37-41.

Louise, D. (2016). Packaging Functional Foods: From Basic Requirements to Nano Perspectives. In:Functional Food Ingredients and Nutraceuticals by Taylor & Francis Group, LLC. pp 409-425.

Maltini, E., Torreggiani, D., Venir, E. and Bertolo, G. (2003). Water activity and the preservation of plant foods. Food Chemistry, 82(1):79-86.

Martirosyan, D. and Singh, J. (2015). A new definition of functional by FFC: what makes a new definition unique? Funct. Foods Health Dis, 5:209-223.

McRae, M. P. (2017). Dietary fiber is beneficial for the prevention of cardiovascular disease: An umbrella review of metanalyses. Journal of Chiropractic Medicine, 16(4):289-299. https://doi.org/10.1016/j.jcm.2017.05.005

Mihindukulasuriya, S.D.F. and Lim, L.T. (2014). Nanotechnology development in food packaging: A review. Trends in Food Science and Technology, 40:149-167.

Naqvi, S., Zhu, C., Farre, G., Ramessar, K., Bassie, L., Breitenbach, J., Perez, C.D., Ros, G., Sandmann, G., Capell, T. and Christou, P. (2009). Transgenic multivitamin corn through biofortification of endosperm with three vitamins representing three distinct metabolic pathways. Proc Natl AcadSci USA, 106:7762-7767.

Nestel, P., Clifton, P., Colquhoun, D., Noakes, M., Mori, T. A., Sullivan, D. and Thomas, B. (2015). Indications for omega-3 long chain polyunsaturated fatty acid in the prevention and treatment of cardiovascular disease. Heart Lung and Circulation, 24(8):769-779. https://doi.org/10.1016/j.hlc.2015.03.020

Nicola, S. and Fontana, E. (2014). Fresh-cut produce quality: Implications for a systems approach. In: Postharvest Handling. A Systems Approach, Ed. R.L.S.W.J. Florkowski, B. Brueckner and S.E. Prussia. Academic Press, San Diego, CA.

Ogbe, R.J., Ochalefu, D.O., Mafulul, S.G. and Olaniru, O.B. (2015). A review on dietary phytosterols: Their occurrence, metabolism and health benefits. Asian Journal of Plant Science and Research, 5(4):10-21.

Pacier, C. and Danik M.M. (2015). Vitamin C: Optimal Dosages, Supplementation and Use in Disease Prevention. Functional Foods in Health and Disease, 5(3):89-107.

Paine, J.A., Shipton, C.A., Chaggar, S., Howells, R.M., Kennedy, M.J., Vernon, G., Wright, S.Y., Hinchliffe, E., Adams, J.L., Silverstone, A.L. and Drake, R. (2005). Improving the nutritional value of golden rice through increased pro-vitamin A content. Nat Biotech, 23:482-487.

Pandey, K.B. and Rizvi, S.I. (2009). Plant polyphenols as dietary antioxidants in human health and disease. Oxidative Medicine and Cellular Longevity, 2(5):270-278. https://doi.org/10.4161/oxim.2.5.9498

Pandey, K.R., Naik, S.R. and Vakil, B.V. (2015). Probiotics, prebiotics and synbiotics- A review. Journal of Food Science and Technology, 52(12):7577-7587. https://doi.org/10.1007/s13197-015-1921-1

Peighambardoust, S.H., Golshan Tafti, A. and Hesari, J. (2011). Application of spray drying for preservation of lactic acid starter cultures: A review. Trends in Food Science and Technology, 22(5):215-224.

Pellegrino, D. (2016). Antioxidants and cardiovascular risk factors. Diseases, 4:11-17. https://doi.org/10.3390/diseases4010011

Perez-Gregorio, R. and Simal-Gandara, J. (2017). A critical review of bioactive food components, and of their functional mechanisms, biological effects and health outcomes. Current Pharmaceutical Design, 23(19):2731-2741. https://doi.org/10.2174/1381612823666170317122913

Perry, A., Rasmussen, H. and Johnson, E. (2009). Xantophyll (lutein, zeaxanthin) content in fruits, vegetables and corn and egg products Journal of Food Composition and Analysis, 22(1):9-15. https://doi.org/10.1016/j.jfca.2008.07.006

Phillips, M.A., Childs, C.E., Calder, P.C. and Rogers, P.J. (2012). Lower omega-3 fatty acid intake and status are associated with poorer cognitive function in older age: A comparison of individuals with and without cognitive impairment and Alzheimer's disease. Nutritional Neuroscience, 15(6):271-277. https://doi.org/10.1179/1476830512Y.0000000026

Poddar, D., Das, S. and Jones, G. (2014). Stability of probiotic *Lactobacillus paracasei* during storage as affected by the drying method. International Dairy Journal, 39(1):1-7.

Ramesh, C.K., Jamuna, K.S. and Raghavendra (2012). Concepts and trends of functional foods: A review. International Journal of Pharmaceutical Research and Development, 4(6):273-290.

Ranjan, S., Dasgupta, N. and Chakraborty, A.R. (2014). Nanoscience and nanotechnologies in food industries: Opportunities and research trends. Journal of Nanoparticle Research, 16(6):1-23.

Rao, Q., Rocca-Smith, J.R. and Labuza, T.P. (2012). Moisture-induced quality changes of hen egg white proteins in a protein/water model system. Journal of Agricultural and Food Chemistry, 60(42):10625-10633.

Ribaya-Mercado, J. D. and Blumberg, J.B. (2004). Lutein and zeaxanthin and their potential roles in disease prevention. Journal of the American College of Nutrition, 23(6):567-587.

Rietjens, I.M.C., Louisse, J. and Beekmann, K. (2017). The potential health effects of dietary phytoestrogens. British Journal of Pharmacology, 174(11):1263-1280. https://doi.org/10.1111/bph.13622

Rizvi, S., Raza, S. T., Ahmed, F., Ahmad, A., Abbas, S. and Mahdi, F. (2014). The role of vitamin E in human health and some diseases. Sultan Qaboos University Medical Journal, 14(2):157-165.

Roberfroid, M.B. (2000). Defining functional foods. Functional foods, 9.

Scholle, J.M., Baker, W.L., Talati, R. and Coleman, C.I. (2009). The effect of adding plant sterols or stanols to statin therapy in hypercholesterolemic patients: Systematic review and metaanalysis. Journal of the American College of Nutrition, 28(5):517-524.

Serino, A. and Salazar, G. (2019). Protective role of polyphenols against vascular inflammation, aging and cardiovascular disease. Nutrients, 11(1):53. https://doi.org/https://doi.org/10.3390/nu11010053

Sharma, M. and Garg, S. (2013). Functional Foods: Marketing 'Health' To Modern India. International Journal of Innovative Research & Development, 2(5):2278-0211

Shukla, V. and Mattoo, A.K. (2010). Potential for engineering horticultural crops with high antioxidant capacity. CAB Revs, 4(066):1-22. Doi: 10.1079/PAVSNNR20094066

Sin, H.P., Liu, D.T. and Lam, D.S. (2013). Lifestyle modification, nutritional and vitamins supplements for age-related macular degeneration. ActaOphthalmologica, 91(1):6-11. https://doi.org/10.1111/j.1755-3768.2011.02357.x

Stavropoulou, E. and Bezirtzoglou, E. (2019). Human microbiota in aging and infection: A review. Critical Reviews in Food Science and Nutrition, 59(4):535-545.

Tager, J. (2014). Nanomaterials in food packaging: FSANZ fails consumers gain. Chain Reaction, 122:16-17.

Tavva, V.S., Kim, Y.H., Kagan, I.A., Dinkins, R.D., Kim, K.H. and Collins, G.B. (2007). Increased α-tocopherol content in soybean seed overexpressing the Perillafrutescens a-tocopherolmethyltransferase gene. Plant Cell Rept, 26:61-70.

Tiihonen, K., Ouwehand, A. C. and Rautonen, N. (2010). Human intestinal microbiota and healthy ageing. Ageing Research Reviews, 9(2):107-116. https://doi.org/10.1016/j.arr.2009.10.004

Tomkin, G. H. and Owens, D. (2012). LDL as a cause of atherosclerosis. The Open Atherosclerosis & Thrombosis Journal, 5:13-21.

Upadhyay, N. and Moudgal, V. (2012). Probiotics: A review. Journal of Clinical Outcomes Management, 19(2):76-84.

Venir, E. and Maltini, E. (2013). Relevance of physical properties in the stability of plant based food products. Indian Journal of Experimental Biology, 51(11):894-904.

Weaver, C.M. (2014). Bioactive foods and ingredients for health. Advances in Nutrition: An International Review Journal, 5(3):306-311. https://doi.org/10.3945/an.113.005124

Weiss, J., Takhistov, P. and McClements, D.J. (2006). Functional materials in food nanotechnology. Journal of Food Science, 71(9):107-116.

Weststrate, J.A., van Poppel, G. and Verschuren, P.M. (2002). Functional foods, trends and future. British Journal of Nutrition, 88(2):233-235.

Yanai, H. (2017). Effects of n-3 polyunsaturated fatty acids on dementia. Journal of Clinical Medicine Research, 9(1):1-9. https://doi.org/10.14740/jocmr2815w

Yusuf, M.A. and Sarin, N.B. (2007). Antioxidant value addition in human diets: genetic transformation of *Brassica juncea* with g-TMT gene for increased α-tocopherol content. Transgenic Res, 16:109-113.

Zaręba, D. and Ziarno, M. (2011). Alternative probiotic vegetable and fruit drinks. Bromatologiai Chemia Toksykologiczna, 2:160-168.

Zawistowska-Rojek, A., Zaręba, T., Mrowka, A. and Tyski, S. (2016). Assessment of the microbiological status of probiotic products. Polish Journal of Microbiology, 65(1):97-104. https://doi.org/10.5604/17331331.1197281

2

Neutraceuticals and Its Impact on Human Health

[1]B.H. Choudhury, [2]P. Koli and **[3]O. Singh**

[1]*Department of BAC, B.N. College of Agriculture, BNCA, AAU Jorhat-13, Assam*
[2]*PAR Division, ICAR-Indian Grassland and Fodder Research Institute, Jhansi Uttar Pradesh*
[3]*Departmentof PHM, RVSKVV Campus, College of Horticulture, Mandsaur Madhya Pradesh*

Introduction

Society is dynamic. The requirement of society varies in course of time. During the life process human being faces lots of health hazards which are mostly due to intake of various foods. So people become cautious about the intake of food and think about the quality of foods which can protect the health of human being. In ancient times people made medicines from plant extracts to protect their health. In 1989, Dr. Stephen coined the term "Nutraceuticals" which is combination of nutrition and pharmaceutical. Nutrition is required for health and pharmaceuticals provide remedy for sickness. So it is a preventive approach. Nowadays, the definition of nutraceuticals has been expanded to include vitamins, minerals, herbs and other botanicals, amino acids, and dietary substance for human use as a supplement diet (Stauffer, 1999).

Nutraceuticals are dietary supplements made only from whole foods to augment health. Currently over 470 nutraceutical and functional food products are available with documented health benefits (Brower, 1999). The patients are much aware about the side effects of chemical agents in short and long term therapy. So to avoid medicine used for small diseases, new therapies came into trend with the help of nutritional approaches i.e. "Nutraceuticals". The use of nutraceuticals becomes easier than pharmaceuticals since nutraceuticals does not require any animal verification for its use in human. Nutraceuticals,

Functional foods and Dietary Supplements are most important concept in food science. They have distinction amongst themselves.

Functional foods *look like food which is modified for greater nutritional value.* As per the definition of American Dietetic Association functional foods "include whole foods and fortified, enriched, or enhanced foods have a potentially beneficial effect on health when consumed as part of a varied diet on a regular basis, at effective levels".

Dietary supplements *look like drugs, made from food, isolated nutrients or food-like substances to augment health. As per the definition of* Dietary Supplement Health and Education Act of 1994 (DSHEA) a dietary supplement is "a product (other than tobacco) that is intended to supplement the diet, which contains one or more of the following dietary ingredients—a vitamin, a mineral, an herb or other botanical, an amino acid, a dietary substance to supplement the diet by increasing the total daily intake, or a concentrate, metabolite, constituent, extract, or combinations of these ingredients; is ingested in pill, capsule, tablet, or liquid form; is not represented for use as a conventional food or as the sole item of a meal or diet; and is labeled as a 'dietary supplement. 'Nowadays peoples are more concern about their health and diet. So peoples are looking for a food which can provide nutrition as well as can provide remedy for diseases. Nutraceuticals are one such type of supplements made from whole food.

Categories of Nutraceuticals

Based on chemical constituents, nutraceuticals can be categorized (Hathcock, 2001) as:

Nutrients

The substances which have established nutritional functions, such as Vitamins like Vitamin A, Vitamin K, Vitamin E, Vitamin C, Vitamin B_1, Vitamin $B_{2,}$ Vitamin $B_{3,}$ Vitamin $B_{6,}$ Folic acid; Minerals like Calcium, Iron, Magnesium, Phosphorous, Chromium, Cobalt, Copper, Iodine, Amino acids and Fatty acids (Allen, 1997). Functions of nutrients are listed in Table1.

Table1: Functions of nutrients

S.N	Nutrients	Functions
1	Vitamin A	Antioxidant, essential for growth and development and in the treatment of certain skin disorders.
2	Vitamin E	Antioxidant, helps form blood cells, muscles, lung and nerve tissue, boost the immune system.
3	Vitamin K	Essential for blood clotting
4	Vitamin C	Antioxidant, essetial for healthy bones, gums, teeth and skin, in wound healing, prevent common cold and attenuate its symptoms.
5	Vitamin B_1	Helps to convert food in to energy, essential in neurologic functions.
6	Vitamin B_2	Helps in energy production and other chemical processes in the body, helps maintain healthy eyes, skin and nerve function.
7	Vitamin B_3	Helps to convert food in to energy and maintain proper brain function.
8	Vitamin B_6	Produce the genetic material of cells, formation of RBCs, maintenance of central nervous system and synthesize amino acids and metabolism of fats, protein and carbohydrates.
9	Folic acid	Produce the genetic materials of cells, in pregnancy for preventing birth defects, RBCs formation, protects against heart disease.
10	Calcium	Bones and teeth and maintaining bone strength important in nerve, muscle and glandular functions.
11	Iron	Energy production, carry and transfer of oxygen to tissues.
12	Magnesium	Healthy nerve and muscle function and bone formation, may help prevent premenstrual syndrome (PMS).
13	Phosphorous	Strong bones and teeth, helps in formation of genetic material, energy production and storage.
14	Chromium	With insulin helps to convert carbohydrates and fats into energy.
15	Cobalt	Essential component of vitamin B_{12}, but ingested cobalt is metabolized in vivo to form the B_{12} coenzymes.
16	Copper	Essential for hemoglobin and collagen production, healthy functioning of the heart, energy production, absorption of iron from digestive tract.
17	Iodine	Essential for proper functioning of the thyroid.

Herbals

Herbs or botanical products as concentrates and extracts like Aloe vera gel, Ginkgo, Garlic, Ginger etc. (Tyler and Foster 1996). Functions of concentrates and extracts are listed in Table 2.

Table 2: Functions of concentrates and extracts

S.N	Herbals (Botanical source)	Functions
1	Aloe vera gel (*Aloe vera* L. N.L. Burm.)	Dilates capillaries, anti-inflammatory, emollient, wound healing properties
2	Chamomile (*Matricaria recutita* L.)	Anti-inflammatory, spasmolytic, antimicrobial, wound Healing
3	Echinacea (*Echinacea purpurea* L.)	Immuno-stimulant, treatment of cold and flu symptoms
4	Ephedra (*Ephedra sinica* Stapf.)	Bronchodilator, vasoconstrictor, reduces bronchial Edema
5	Evening primrose oil (*Oenothera biennis* L.)	Dietary supplement of linoleic acid, treatment of atopic eczema
6	Feverfew (*Tanacetum parthenium* L.)	Treatment of headache, fever and menstrual problem, severity and duration of migraine headaches.
7	Garlic (*Allium sativum* L.)	Antibacterial, antifungal, antithrombotic, hypotensive anti-inflammatory
8	Ginger (*Zingiber officinale* Rosc.)	Carminative, antiemetic, cholagogue, positive inotropic
9	Ginseng (*Panax ginseng*)	Adaptogen
10	Ginkgo (*Ginkgo biloba* L.)	Vasodilation, increased peripheral blood flow, treatment of post thrombotic syndrome
11	Goldenseal (*Hydrastisca nadensis* L.)	Antimicrobial, astringent, antihemorrhagic, treatment of mucosal inflammation, dyspepsia, gastritis
12	Horehound (*Marrubium vulgare* L.)	Expectorant, antitussive, choleretic
13	Licorice (*Glycyrrhiza glabra* L.)	Expectorant, secretolytic, treatment of peptic ulcer
14	Melissa (*Melissa officinalis* L.)	Topical antibacterial and antiviral
15	Plantago seed (*Plantagoarenaria* Waldst)	Cathartic
16	St. John's wort (*Hypericum perforatum* L.)	Anxiolytic, anti-inflammatory, antidepressant, monoamine oxidase inhibitor
17	Valerian (*Valeriana officinalis* L.)	Spasmolytic, mild sedative, sleep aid
18	Willow bark (*Salix alba* L.)	Anti-inflammatory, analgesic, antipyretic, astringent, treatment of rheumatic and arthritic

Dietary Supplement

Dietary supplements contain a dietary ingredient added to the foods we eat. Examples of dietary supplements are black cohosh for menopausal symptoms, ginkgo biloba for memory loss, and glucosamine/chondroitin for arthritis. They also serve specific functions such as sports nutrition, weight-loss supplements and meal replacements. Supplements may incorporate vitamins, minerals, herbs or other botanicals, amino acids, enzymes, organ tissues, gland extracts, or other dietary substances. They are available in different forms like tablets, capsules, liquids, powders, extracts, and concentrates (www.nccam.nih.gov.in). Functions of dietary supplements are listed in Table 3.

Table 3: Functions of dietary supplements

S.N	Dietary supplements	Functions
1	Ketogenic diets	Comprised of foods with high fat and low protein and carbohydrate content, have been reported to improve seizure control. However, these diets are widely acknowledged to be unpalatable (Chapman and Giles, 1997).
2	Minimally refined grains	Cereals and grains fortified with calcium may reduce the incidence of diabetes (Salmeron, et al. 1997) and prevents gastrointestinal cancers (Slavin et al., 1997).
3	Phytoestrogens	Found in soya flour and linseeds and have been documented to enhance estrogens levels when hormonal levels are low. This action may prevent against both hot flushes and breast cancer (Chang, R., 1996).
4	Several species of edible mushrooms	*Tonnage, Lentinus, Pleurotus, Auricularia, Flammulina, Tremella, Hericium* and *Grifola* have varying degrees of immunomodulatory, lipid lowering and antitumor without any significant toxicity.
5	Glucosamine sulfate and chondroitin sulfate	They are effective and safer to alleviate symptoms of osteoarthritis (Reyes et al., 2000).
6	Peptides/Hydrolysates	Found in casein and whey protein and have A.C.E. inhibitor activity (Korhonen et al., 2000). Buckwheat proteins used as flour reduces cholesterol, hypertension; improve constipation and obesity by acting similar to dietary fibers and interrupting the in-vivo metabolism (Li and Zhang, 2001).
7	Dairy foods	Containing friendly or probiotic bacteria claimed to promote gut health. Bio yoghurts containing *Lactobacillus acidophilus* and *Bifidobacteria* lead the sector (Morelli, 2002).

Traditional and Non- Traditional Nutraceuticals

Lots of nutraceutical foods are available in the market; they fall under the category of traditional foods and non- traditional foods.

Traditional Nutraceuticals

Traditional Nutraceuticals are foods in which no change is made. It is natural, whole foods with new information about their potential health qualities. Many fruits, vegetables, grains, fish, dairy and meat products contain several natural components which provides benefits besides basic nutrition, such as lycopene in tomatoes, omega-3 fatty acids in salmon or saponins in soy. Even tea and chocolate have some health-benefiting attributes. Tomatoes and salmon contain benefits besides basic nutrition - lycopene and omega-3 fatty acids, respectively.

Non-traditional Nutraceuticals

These are the outcome of agricultural breeding or addition of nutrients such as orange juice fortified with calcium; cereals with added vitamins or minerals and flour with added folic acid, technologically improved crops are nontraditional nutraceuticals.

Growing awareness about nutrition and diet created favorable market conditions for Nutraceutical industry in India. Nutraceutical industry of India has a chance to boost up due to some factors like qualified human resources, world class R & D facilities and varied raw material-aspects. The Indian nutritional market is estimated to be USD 1 Billion. The global market is growing at the compound annual growth rate (CAGR) of 7% while Indian market growing at the CAGR of 18% for the last three years. However, market potential in India is two to four times than the current market which is between USD 2 to USD 4 billion. The Indian nutraceutical market is dominated primarily by pharmaceuticals and FMCG companies with very few pure play nutraceutical companies.

Conclusion

Many nutraceuticals, functional foods and naturally occurring compounds are extremely active, have profound effect on cell metabolism with little adverse effect. It is natural that people giving emphasis on prevention of diseases to remain healthy. Nutraceuticals provide an advantage over the synthetic drugs developed by the pharmaceuticals industry since it provide nutrition as well as prevention in several diseases.

References

Allen, L.V. (1997). Nutritional Products, In: Covington TR, Berardi RR, Young LL, et al. Editors. Handbook of Non-prescription Drugs. Washington DC: American Pharmaceutical Association.

Brower, V. (1999). Nutraceuticals: poised for a healthy slice of the healthcare market. Nat. Biotechnol.16: 728-731.

Chang, R. (1996). Functional properties of edible mushrooms. Nutr Rev.54:91-93.

Chapman, D.P. and Giles, W.H. (1997). Pharmacologic and dietary therapies in epilepsy: Conventional Treatments and Recent Advances. South Med. J. 90: 471-80.

Hathcock, J. (2001). Dietary supplements: How they are used and regulated. J. Nutrition. 131: 1114-1117.

Korhonen, H., Marnila, P. and Gill, H.S. (2000). Bovine milk antibodies for health.Br. J.Nutr.84:135-46.

Li, S.Q. and Zhang, Q.H. (2001). Advances in the development of functional foods from buckwheat. Crit Rev Food SciNutr.4: 451-464.

Morelli, L. (2002). Probiotics: clinics and/or nutrition. Dig Liver Dis. 34:8-11.

Reyes, G.C., Koda, R.T., and Lein, E.J. (2000). Glucosamine and Chondroitin sulfates in the treatment of osteoarthritis: a survey. Prog Drug Res.55: 81-103.

Salmeron, J., Manson, J.E., Stampfer, M.J., et al. (1997). Dietary fiber, glycemic load, and risk of noninsulin-dependent diabetes mellitus in women. JAMA.277: 472-7.

Slavin, J., Jacobs, D. and Marquart, L. (1997).Whole grain consumption and chronic disease: protective mechanisms. Nutr Cancer. 27: 14-21.

Stauffer, J.E. (1999): Nutraceuticals. Cereals Food World 44(2): 115-116.

Tyler, V.E. and Foster, F. (1996). Herbs and phytochemicals, In: Covington TR, Berardi RR, Young LL et al. editors. Handbook of Non-prescription Drugs. Washington DC: American Pharmaceutical Association.

www.nccam.nih.gov.in National Institutes of Health National Center for Complementary and Alternative Medicine.

3

Significance of Bioactive Compounds from Natural and Dietary Interventions in Cancer Chemoprevention

[1]Shiv Kumar Dubey, [2]Kanchan Gairola, [1]Shriya Gururani and [2]Rakesh Sharma

[1]Department of Biochemistry, College of Basic Sciences and Humanities G.B. Pant University of Agriculture and Technology Pantnagar-263145 Uttarakhand
[2]Department of Food Science and Technology Dr. Y.S. Parmar University of Horticulture and Forestry, Nauni Solan-173230, Himachal Pradesh

Introduction

Our traditional knowledge of medicine has been one of the strongest areas of research because of our rich heritage in herbal plants & great biodiversity in our fauna & flora. Consumption of plant foods and the structurally diverse natural products provided by them have long been associated with health-protective and therapeutic effects. Natural products in plant foods have been shown to have pleiotropic therapeutic and chemopreventive properties. These bioactive products are known to potentiate the efficacy of other drugs and have synergistic effects too. Most of the prescription drugs discovered so far is from plant sources and 75% drugs had been claimed in folklore claims and traditional medicinal system (Reynold, 1991; Benowitz, 1996). The traditional Indian medicinal system utilizes multicomponent formulations of plant extracts which may either act synergistically or avoid many side effects and might have added up trace element through combination of different plant extracts which could have additive effects. However the effort to utilize traditional knowledge in the light of modern scientific findings is lacking in India. Accumulating evidences as well as in-vitro studies implicates inverse relationship between regular consumption of fruits vegetable and the risk of specific cancer. Diet derived nutritional compound mediated interventions in the process of tumorigenesis may be promising chemopreventive strategy with minimal toxicities to the physiological system.

Mechanism of Carcinogenesis

Cancer, a disease characterized by uncontrolled growth of cells, still remains a global problem to be interrogated because of its high incidence, recurrence and mortality. Carcinogenesis is the multistep process involves in transformation of normal cell into cancer cell due to some alteration and mutation. Three basic steps which are involved in carcinogenesis are initiation, promotion and progression. Initiation is the irreversible process that leads to transformation of normal cell so that these cell competent to form tumors. Promotion generally refers to the increase in number of initiated cells due to increase in proliferation and progression is the last step of neoplastic transformation that leads to malignant changes in tumor cells.

Most of traditional chemotherapeutic strategies implicated for cancer therapy involve targeting proliferating cells manifested in neoplastic transformations. Therapies aimed at targeting these transformed cells have not been shown efficient because of recurrence, drug resistance, poor prognosis and there have been substantial toxicity to normal tissue too. The initiation and re-emergence of cancer has been attributed to a subpopulation of few mutated and transformed cells in tumor population, called tumor initiating cells (TICs) or cancer stem cells (CSC), which acquire capacity of self-renewal and differentiation. Cancer stem cells are known to resist the traditional methods of chemo and radio therapy (Ke.Chen et al., 2013; Lu Han et al., 2013).

Cancer Chemoprevention

Chemoprevention is the utility of chemicals, drugs, food ingredients, vitamin or other substance to intervene with the process of disease (Sporn and Suh, 2000). Cancer chemoprevention is defined as use of agents that can delay or disturb process of tumorigenesis. Chemopreventive agents help in blocking or reversing the premalignant stages of carcinogenesis (Kwon et al., 2007). Therefore, capabilities of a single chemopreventive agent to cure cancer should be acknowledged by the results of combination of different intracellular effects.

Many chemo preventive agents block or delay the progression of transformed cell by modulating cell proliferation or differentiation. According to convention classification proposed by Wattenberg (Wattenberg, 1981), Chemo preventive agents are broadly divided into two categories i.e. 1) Blocking agent and 2) Suppressing agent. Blocking agent in carcinogenic process restricts mutagenic initiation and avert carcinogen from reaching target site, undergoing metabolic activation and from interacting with crucial cellular macromolecules (DNA,

RNA or Protein). Suppressing agent hinder the deadly cancerous transformation of initiated cell in either the progression or promotion stage. In another word Suppressing agent are those agents that inhibit further progression of lesion that has been already established.

Mechanism of blocking agent	Mechanism of suppressing agent
Free radical's scavenging	Alteration in gene expression
Show Antioxidant activity	Inhibition of cell proliferation clonal expansion.
Inhibition or restriction of Phase I drug metabolizing enzyme	Induction of apoptosis in preneoplastic lesions.
Induction of phase II drug metabolizing enzyme	Induction of terminal differentiation, senescence.
Induction of DNA repair & Carcinogen uptake blockade	Signal transduction modulation

Natural Compound in Cancer Chemoprevention

Consumption of plant foods and the structurally diverse natural products have long been associated with health-protective and therapeutic effects. Natural products obtained from plant foods and dietary intervention has been documented to show cancer chemo preventive properties. The bioactive natural compounds components and their derivatives like Curcumin, Gingerol, Epigallocatechin -3-gallate, Lycopene, Genistein, Sulforaphane etc. obtained from herbs and dietary components having cancer chemo preventive properties, are known to potentiate the efficacy of other drugs and have synergistic effects too.

Table 1: Natural dietary compound and the chemo preventive agents with their chemical structure

Dietary compound	Chemo preventive phytochemicals	Chemical Structure
Turmeric	Curcumin	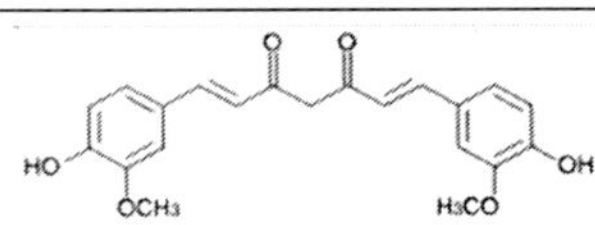
Ginger	Gingerol	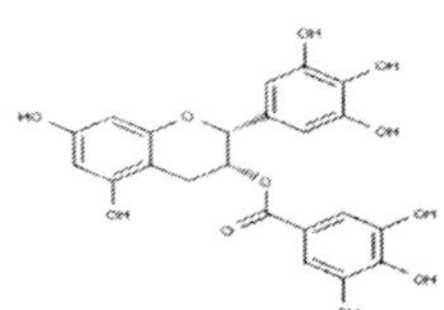
Green tea	Epigallocatechin -3-gallate	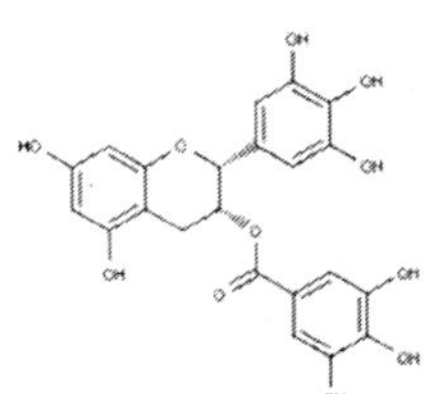

Dietary compound	Chemo preventive phytochemicals	Chemical Structure
Soybean	Genistein	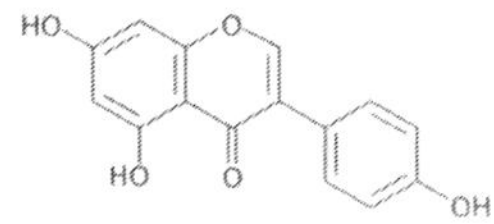
Broccoli	Sulforaphane	
Cabbage	Indole -3- carbinol	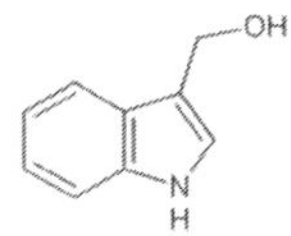

Curcumin

Turmeric commonly used in Indian cuisines and curries extracted from rhizome of *Curcumin longa* which belong to zingiberaceae family has immense medicinal properties in Indian traditional medicinal system. Curcumin, bioactive principle of turmeric, have pleiotropic medicinal properties. Curcumin has shown chemopreventive properties against wide variety of cancers. Curcumin block phosphorylation & suppress TNF alpha and induce nuclear translocation and DNA binding of NF-kB in human myeloid leukemia cell line (Singh and Aggrawal, 1995). In human multiple mylenoma cells (Bharti et al, 2003) and murine melanoma cells (Philip and Kundu, 2003). Jaiswal and co-workers have proposed that caspase-3-mediated cleavage of beta – catenin was induced by curcumin that leads to inactivation of wnt/ beta catenin signaling in HCT116 (Jaiswal et al, 2002). Wnt/beta catenin is important signaling pathway irrregulation of which leads to formation of cancer stem cells which are implicated for initiation and recurrence of the disease. It was further supported by the work of Park et.al who suggested that Curcumin decreased beta catenin /TCF transcription activity in all tested cancer cell line including gastric, colon and intestinal cancer cell which was applied to reduce the amount of nuclear beta catenin and TCF-4 proteins (Park et al, 2005). Curcumin was also able to vitiate response of Beta catenin to wnt-3a in colon cancer cell through down regulating p300 (positive regulator of Wnt pathway) (Ryu et al, 2008). In a recent case study, Kakarala and others express that curcumin was able to target breast stem/progenitor cells as evidenced by suppressed mammosphere formation and a decrease in the percent of ALDH-positive cell whereas curcumin had less impact on differentiated cells (Kakarala et al, 2010).

Gingerol

Gingerol, a phenolic compound from ginger shows anti-tumor and anti-metastatic activity against various tumor types including breast cancer. Gingerol is involved in inducing cell cycle arrest, elevating intracellular Ca_2^+ and caspase dependent apoptosis. Recent studies show that 6- gingerol significantly reduced cell viability of human colon cancer cells. Gingerol elevate intracellular reactive oxygen species (ROS) and phosphorylation level of p53. These studies suggest that gingerol could be beneficial to colon cancer treatment (Ching-Bin Lin et al, 2012).

Epigallocatechin -3-gallate (EGCG)

EGCG, obtained from green tea, is known to play a major role in anticancer effect by inhibiting NF-kB activity, MAPK pathway and activator protein -1 (AP-1) activity. EGCG shows its cytotoxic effect on breast cancer cell and decrease cell number control in the ER positive cell lines MCF-7 and BT474. EGCG reduce invasiveness and tumorigenic proliferation in cancer breast cell by blocking Wnt signaling by stabilizing mRNA of HBP1 which is a suppressor of Wnt signaling. EGCG treatment could cause inhibition in cell growth, by arresting cell cycle at Go/G1 phase in human epidermoid carcinoma (A431) but not in normal human epidermoid carcinoma keratinocytes (NHEK) (Ahmad et al, 2000).

Genistein

Genistein an isoflavanone that is present in soybean is considered as a putative preventer of breast and prostate cancer. Genistein as a chemopreventive agent has Cyclo-oxygenases -2 (COX-2, also known as prostaglandin H synthase) pathway as its potential target. COX -2 is an inducible enzyme which is commonly expressed in inflammatory environment under modulation by cytokines, mitogens, growth factors and tumor promoters. Genistein can also help in inhibiting the expression level of COX-2 in vivo and in vitro which in turn contributed ultimately to the chemopreventive effect of Genistein (Aggarwal and Shishodia, 2006).

Sulforaphane

Sulforaphane, produced by the hydrolytic conversion of glucoraphanin after ingestion of cruciferous vegetables, particularly broccoli and broccoli sprouts, has been extensively studied due to its apparent health-promoting properties in disease and limited toxicity in normal tissue. Sulforaphane (SFN) not only effectively involved in reducing cancer that is chemically induced but also involved in inhibition of growth of the existing tumors. Kallifatidis et.al, suggested that sulforaphane by interrupting the TRAIL (tumor necrosis factor related apoptosis inducing ligand) activated NF-KB signaling could abolish the resistance of pancreatic tumor initiating cells (TICs) to TRAIL (Kallifatidis et al, 2009).

Indole -3- carbinol

Indole-3-carbinol present in various vegetables like cabbage, Brussels sprouts etc. is considered as a cancer preventing agent. Indole-3-carbinol activates the stress induced MAP kinases p38 and c-jun N terminal kinase (JNK) in prostate cancer cells and also to constitutively active STAT3, a transcription factor, in pancreatic carcinoma cell. Indole-3-carbinol inhibit invasion and migration of breast cancer cell by inflect the expression of signaling proteins related with invasion and migration including E-cadherin, catenin, BRAC1, the chemokine receptor CXCR4 and the matrix metalloproteinase (MMP)-9 (Rahman et al, 2006).

Conclusion

Taking consideration of financial burden and toxicity to other physiological system by chemotherapeutic drugs for cancer treatment, diet derived agents might be novel strategy for prevention of cancer. The advantage of these dietary/nutritional interventions is that their target proteins are already known. Neutraceuticals are well tolerated, found in many food products that can be added in the diet and can be taken on long term. Thus, diet derived nutritional interventions mediated events in the process of tumorigenesis may be promising chemopreventive/therapeutic strategies with minimal toxicities to the physiological system.

Utilizing nutritional/dietary interventional in traditional medicinal system have advantage over modern medicine that nutritional interventions are very well tolerated, found in many food products that can be added in the diet and can be taken on long term with minimal toxicity to living system. Recent findings with modern medicine are that multitargeted therapy works better than monotherapy. Modern medicine uses single, pure, discrete chemical entities which generally have substantial toxicity to the body system. A combination of the traditional medicine and modern therapy is probably the best answer.

References

Aggarwal BB, Shishodia S. Molecular targets of dietary agents for prevention and therapy of cancer. Biochemical pharmacology; 71:1397-1421(2006).

Ahmad, N., Gupta, S. & Mukhtar, H. Green tea polyphenol epigallocatechin-3-gallate differentially modulates nuclearfactor κB in cancer cells versus normal cells. Arch. Biochem. Biophys. 376, 338–346 (2000).

Benowitz S. New technologies and approaches spur industry interest in plant derived drugs. The Scientist 10 (3): 1, 6 (1996).

Bharti, A. C., Donato, N., Singh, S. & Aggarwal, B. B. Curcumin (diferuloylmethane) down-regulates the constitutive activation of nuclear factor-κ B and IκBα kinasein human multiple myeloma cells, leading to suppression of proliferation and induction of apoptosis. Blood 101, 1053–1062 (2003).

Ching-Bin Lin, Chun-Che Lin, and Gregory J. Tsay2, 4 6-Gingerol Inhibits Growth of Colon Cancer Cell Lo Vo via Induction of G2/M Arrest. Hindawi Evidence-Based Complementary and Alternative Medicine Volume 2012,

Jaiswal AS, Marlow BP, Gupta N, Narayan S. Beta-catenin-mediated transactivation and cell–cell adhesion pathways are important in curcumin (diferuloylmethane)- induced growth arrest and apoptosis in colon cancer cells. Oncogene; 21:8414–27 (2002).

Kakarala M, Brenner DE, Korkaya H, et al. Targeting breast stem cells with the cancer preventive compounds curcumin and piperine. Breast Cancer Res Treat; 122:777–85(2010).

Kallifatidis G, Rausch V, Baumann B, et al. Sulforaphane targets pancreatic tumor initiating cells by NF-kappaB-induced antiapoptotic signalling. Gut; 58:949–63(2009).

Ke Chen, Ying-hui Huang and Ji-long Chen, Understanding and targeting cancer stem cells: therapeutic implications and challenges, Acta Pharmacologica Sinica 34: 732–740 (20130.

Kwon KH, Barve A, Yu S, Huang MT, Kong AN. Cancer chemoprevention by phytochemicals: potential molecular targets, biomarkers and animal models. Acta Pharmacol Sin. Sep; 28(9):1409-21 (2007).

L.W. Wattenberg, Inhibitors of chemical carcinogens, in: J.H. Burchenal, H.F. Oettgen (Eds.), Cancer: Achievements, Challenges and Prospects for the 1980s, Grune and Stratton, NewYork, NY, 1981, pp. 517–540.

Lu Han, Sanjun Shi, Tao Gong, Zhirong Zhang, Xun Sun,Cancer stem cells: therapeutic implications and perspectives in cancer therapy, Acta Pharmaceutica Sinica B, 3(2), 65-75, (2013).

Park CH, Hahm ER, Park S, Kim HK, Yang CH. The inhibitory mechanism of curcumin and its derivative against beta-catenin/Tcf signaling. FEBS Lett; 579:2965–71 (2005).

Philip, S. & Kundu, G. C. Osteopontin induces nuclear factor κB-mediated promatrix metalloproteinase-2 activation through IκBα/IKK signaling pathways, and curcumin (diferulolylmethane) down-regulates these pathways. J. Biol. Chem. 278, 14487–14497 (2003).

Rahman KM, Sarkar FH, Banerjee S, Wang Z, Liao DJ, Hong X, et al. Therapeutic intervention of experimental breast cancer bone metastasis by indole-3- carbinol in SCID-human mouse model. Mol. Cancer Ther. 5:2747–2756 (2006).

Reynold T, Tropical rain forest conservation tied to drug development. J. Natl. Cancer Inst. May 83(9), 594-596 (1991).

Ryu MJ, Cho M, Song JY, et al. Natural derivatives of curcumin attenuates the Wnt/beta-catenin pathway through down-regulation of the transcriptional coactivator p300. Biochem Biophys Res Commun; 377:1304–8 (2008).

Singh, S. & Aggarwal, B. B. Activation of transcription factor NF-κ B is suppressed by curcumin (diferuloylmethane). J. Biol. Chem. 270, 24995–25000 (1995).

Sporn MB, Suh N. Chemoprevention of cancer. Carcinogenesis, 21:525–30 (2000).

4

Processing of Functional Food Ingredients

[1]Priti Khemariya and [2]Tanweer Alam

[1]Indian Institute of Packaging, Delhi-110092
[2]Indian Institute of Packaging, Mumbai, Maharashtra

Introduction

Consumers are nowadays very much concerned about the food they purchase and consume. Consumer wants the food containing antioxidant properties or can say functional food in safe and stable forms, so that the availability of daily requirements of functional food ingredient must be ensured for normal growth and development of the body. The functional food that can provide a health benefit such as increased dietary fibre, vitamins or minerals to the body. The incorporation of antioxidant compounds acts as an efficient way to fight against reactive oxygen species (ROS) generated in our body. As it is already known that the ROS causes oxidative stress and acts as a main trigger for different life threatening chronic diseases like cancer, diabetes, cardiovascular disease, aging etc. Therefore, seeing the consumers' demand for value added products with bioactive ingredients, all small and large scale food industries are taking initiatives to process different types of food (fruits, vegetables, dairy, sea food, animal, microbial products etc.) to extract and purify its active fraction of functional ingredients to valorize different value added products. Functional ingredients may also be derived from different food wastes generated during processing of food. This practice provides additional economic benefits to food businesses. Food industries are also giving importance to different processing methods to extract high quality of functional food ingredients in large quantity because processing method affects the concentration of extracted bioactive compounds. The simple addition of functional ingredients may cause an unwanted and negative change of sensory and texture of food. Therefore, using novel processing technologies such as microencapsulation, the negative changes could be minimized or avoided because these technologies could prevent the inactivation of bioactive ingredients. The food industries either add functional ingredient in raw materials such as addition of nutraceutical in vegetables or they add these in processed food products which is ready to eat such as addition of omega 3 in fruit juices (Wang and Bohn 2012). Keeping this point, the present chapter has been designed to overview different types of extraction methods for processing of functional food ingredients in purified form so that these can be incorporated in manufactured foods to make value added products of health benefits. This chapter will also be discussed on the stability of functional ingredients during different thermal and non-thermal food processing methods used in manufacturing various processed food products for consumption at home or by food manufacturing industries.

Preparation of Functional Foods

There are three types of functional foods (Fig. 1) first is functional packaged processed food (bread, yogurt, cheese, margarines etc.), second is functional packaged beverage (energy drink, probiotic drink etc.) and third is functional fresh food (egg, vegetables etc.). Following is the detailed procedure for preparation of functional foods (Fig. 2):

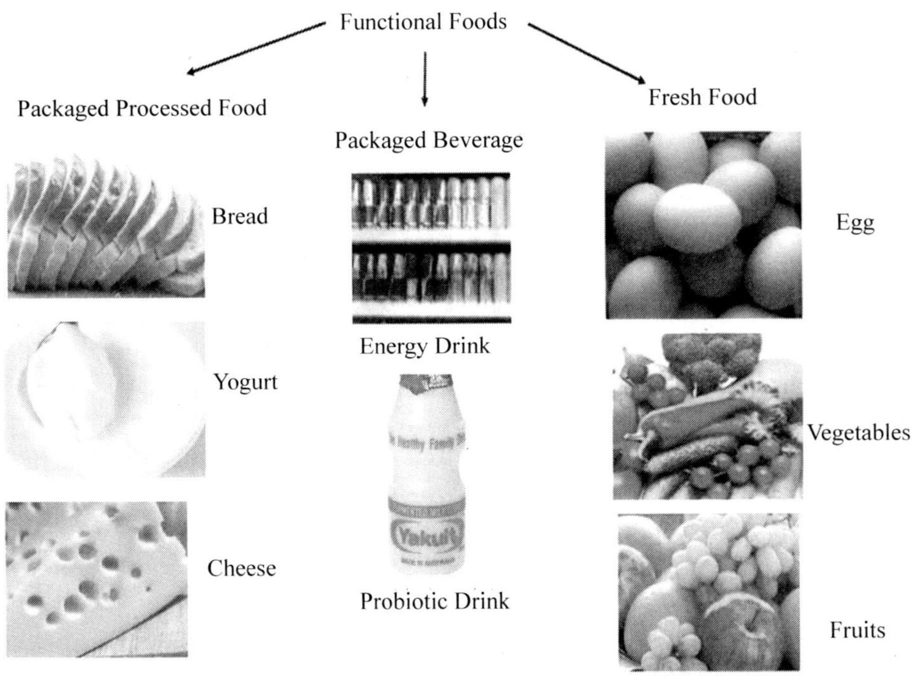

Fig. 1: Category of Funcitonal Food

Selection of Sources

↓

Extraction of Functional Ingredients

↓

Purification of Extracted Functional Food Ingredients

↓

Packaging of Functional Food Ingredients

↓

Preservation of Functional Food Ingredients

↓

Storage of Extracted Functional Food Ingredients

↓

Application of Functional Food Ingredients in Foods to make Functional Foods

Fig. 2: Processing of Functional Food Ingredients by Manufacturing

Selection of Sources

Selection of sources is the first step to extract out the functional ingredients. Functional foods are manufactured by food manufacturers by addition of desired quantity of functional ingredients extracted from different sources like plants (cereals, oilseeds, legumes, vegetables, fruits, herbs, spices) animals (fish, poultry, meat, egg, milk) and microorganisms (yeast, *Bifidobacterium*sp. *Lactobacillus* sp.) (Fig. 3).

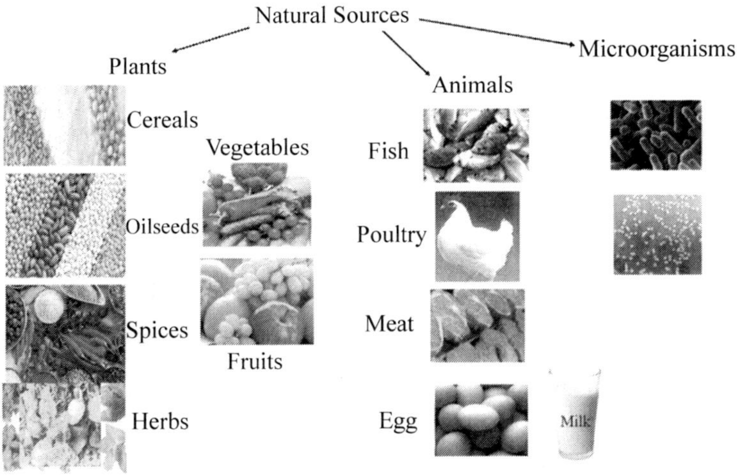

Fig. 3: Availability of natural sources for extraction of functional food ingredients

Extraction of Functional Ingredients

Extraction is the most important step to recover the functional ingredients from the samples. Extraction and isolation of functional ingredients is a time consuming process. Therefore, selection and application of suitable extraction process is a critical process. There are several types of extraction methods described in different literatures.

Types of Extraction Method

Extraction methods can be categorized in to two following parts (Fig. 4):

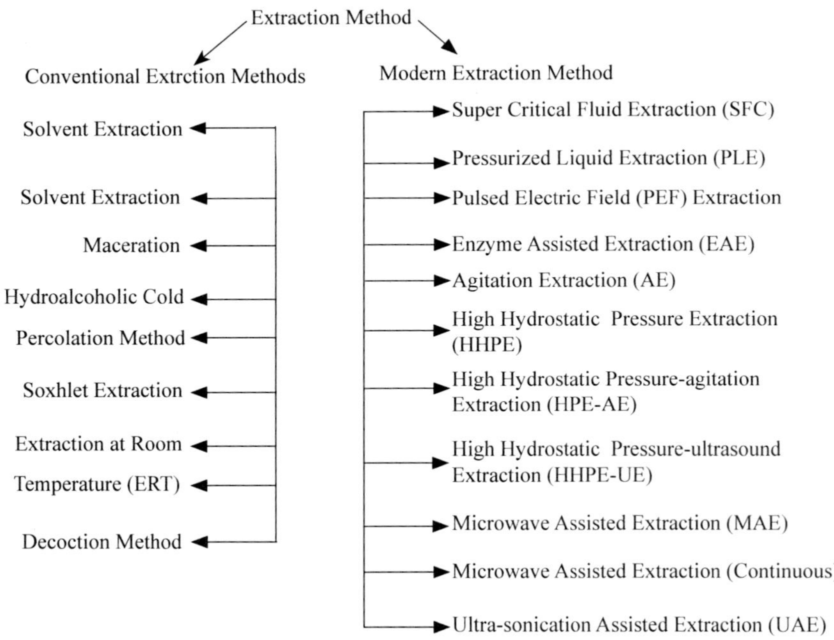

Fig. 4: Different methods of Extraction of Food Functil Ingredients

Conventional Extraction Methods

Solvent Extraction

It is most widely used extraction method of functional ingredients from the samples. Ethanol and methanol are commonly used solvents for extraction of functional ingredients. This method involves the penetration of solvent inside the solid matrix of the sample followed by solutes in the matrix is dissolved in solvent and it diffused out of matrix that are collected as fraction of functional ingredients (Zang et al. 2018).

There are several factors affecting the extent of extraction.

- The level of diffusion of solvent inside the solid matrix and the extent of solubility of solvent in matrix are major determining factors to extract maximum functional fraction. The more level of diffusion and solubility the greater the extraction of functional ingredients will be occurred. In other terms, one should use the solvent of similar polarity of the solute of the solid matrix of the sample because solvent has the tendency of intermiscibility i.e. like dissolves like. Therefore, the selection of suitable solvent is an important point of consideration.
- Similarly, the temperature used for extraction process also affects the amount of recovery of functional ingredients. If temperature is higher the amount of extraction will be higher as temperature enhances the diffusion and solubility of the solvent in solid matrix of the sample. But precaution should be taken while considering the temperature because if temperature is more than the requirement it will evaporate the solvent and also can decompose the heat sensitive components in the fraction to be extracted.
- The time taken for solvent treatment with sample is also another parameter. The duration of exposure of solvent increases the extraction of the functional ingredients. But once the equilibrium of solute is established inside and outside the solid matrix, no extraction takes place even after increasing time exposure (Zang et al. 2018).

Accelerated Solvent Extraction

This technique uses solvents at elevated temperatures and pressures to maintain the solvent in liquid form during the extraction process. The elevated temperature solubilizes the solvent and thus the diffusion rate increases. The higher temperature also reduces the viscosity and thus the solvent can easily penetrate the pores of the matrix of the samples. The elevated pressure enables close contact of samples with the solvent (Mottaleb and Abdul 2012). Benefits from the technique is as follows:

- This technique can extract functional ingredients for sample sizes 1-100g in minutes.
- This technique utilizes less time and solvent for extraction.

Maceration

In this technique, small sized sample pieces is immersed in suitable solvent like water, alcohol oil etc. in stoppered container at room temperature for a certain time period with frequent agitation. After that, solvent is drained off or centrifuged to remove solid part of the sample. Following consideration are needed while doing maceration of sample by this technique:

- The size of sample matters the extraction process. Small sized pieces of sample increase the contact between sample and solvent. However, solvent cannot penetrate the big sized samples. When the sample is taken in powdered form, the volatile functional ingredients may get lost.
- The selection of solvent is also important. Commonly, alcohol is widely used solvent. Alcohol can extract both hydrophilic and lipophilic fraction from the sample. Oil such as vegetable oil, olive oil, seed oil etc. is also used but this extract only lipophilic fraction of the functional ingredients. However, water can extract only hydrophilic fraction.
- Temperature increases diffusion coefficient and decreases viscosity.
- Solid solvent ratio is very important as yield decreases with increasing quantity of sample with constant quantity of solvent.
- Steeping and swelling of the sample in solvent increases capillary dilation and diffusion rate of extract (Zang et al. 2018).

Hydroalcoholic Cold Percolation Method

In this technique, an amount of dried powdered sample is immersed in 100 mL of petroleum ether in a conical flask, plugged with cotton wool and then kept on a rotary shaker at 120 rpm for 24 h of time followed by it is filtered using 8 layers of muslin cloth. The filtrate is then centrifuged at 5 000 rpm for 15 min and the supernatant is collected and air dried. The dried content is then taken in suitable solvent and is stored in air tight bottles at 4 °C.

Soxhlet Extraction

The Soxhlet extraction is an automatic continuous extraction method. It is performed using Soxhlet apparatus for 3 h for all used solvents. It utilizes the principle of reflux and siphoning to continuously extract using fresh solvent. Soxhlet Extractor is consisted of following parts:

- Percolator (Boiler and Reflux)-It circulates the solvents.
- Thimble of thick filter paper- It retains sample to be extracted out.
- Siphon-It periodically empties the thimble.

The high temperature and long extraction time in the Soxhlet extraction may cause thermal degradation. This method is widely used when the desired compound has a limited solubility in the particular solvent and impurities are less soluble in the solvent (Azwanlda 2015).

Extraction at Room Temperature (ERT)

It is carried out in flask on a shaker for 24 h at room temperature. Flask was covered with aluminum foil to prevent any evaporation.

Decoction Method

In this method, a certain quantity of dried powdered sample is extracted with 100 mL of deionized water at 100 °C for 30 min in a water bath followed by it is filtered with eight layers of muslin cloth and centrifuged at 5 000 rpm for 10 min. The supernatant is collected and the solvent is evaporated to dryness is stored in air tight bottle at 4 °C. Water soluble and thermolabile constituents can be extracted by this method (Mittal et al. 2012).

Modern Extraction Method

Super Critical Fluid Extraction (SFC)

In this technique, supercritical gases such as CO_2, N_2, CH_4, SO_2, NH_3, SF_6 ethane, ethylene, nitrous oxide, propane and propylene are used to extract active functional ingredients from samples. For extraction purpose, sample is kept with a gas in container under controlled temperature and pressure. The active ingredients which dissolved in the gas separate or show mass transfer when both temperature and pressure are lower. Supercritical carbon dioxide (S-CO_2) is widely used. It has low critical temperature i.e. 31 °C, selectivity, inertness, low cost and non-toxicity. It is also capable to extract thermally labile compounds. It has low polarity which makes it ideal for the extraction of non-polar natural products (lipid and volatile oil) (Sánchez-Camargo et al, 2014).

Advantages

- Super critical fluid can be recovered from sample extract due to its volatile nature.
- Super critical fluid is harmless fluid
- Components with high boiling point can be extracted out at low temperature
- Thermo-labile compounds can be separated out easily because low temperature is used for extraction.

Disadvantages

- Instrument is costly.
- High pressure is required.

Pressurized Liquid Extraction (PLE)

In this technique, a high pressure is applied in extraction. High pressure keeps solvents in a liquid state above their boiling point resulting in a high solubility and high diffusion rate of lipid solutes in the solvent, and a high penetration of the solvent in the matrix (Fig.5) (Alvarez-Rivera et al. 2020).

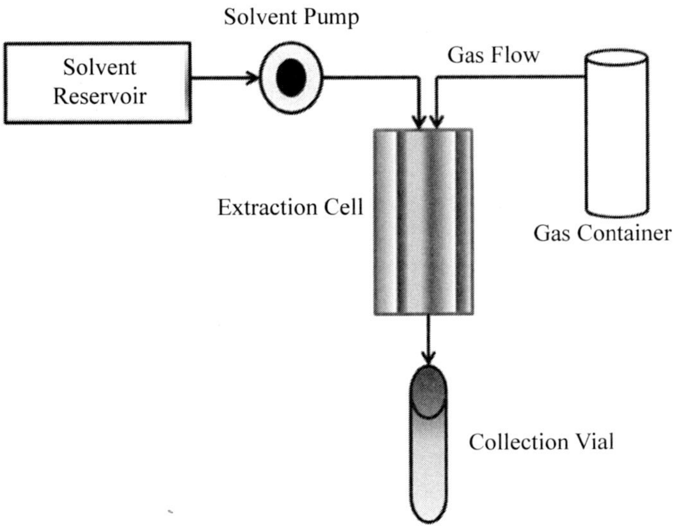

Fig. 5: Diagrammatic representation of Pressurized Liquid Extraction (PLE) Instrumentaion

Pulsed Electric Field (PEF) Extraction

In this technique pulse electric field is applied for extraction. It is a non-thermal extraction process. Hence, thermolabile extracts can be easily extracted out. Pulsed electric field extraction significantly increases the extraction yield and decreased the extraction time because it can increase mass transfer by destroying membrane structures. While using this technique, field strength, specific energy input, pulse number and treatment temperature are important criteria (Fig.6) (Carullo et al. 2020).

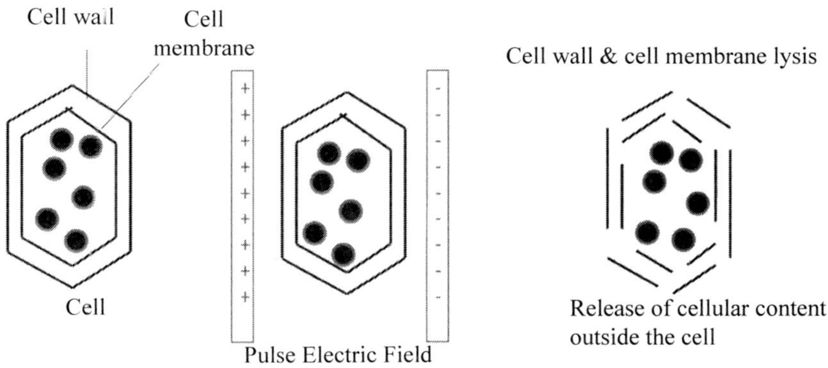

Fig. 6: Effect of Pulsed Electric Field (PEF) on cell

Enzyme Assisted Extraction (EAE)

This technique utilizes various hydrolytic enzymes such as Cellulose, hemicellulase, glucose oxidase, amyloglucosidase, α-amylase and pectinase to break down sample components (such as cell membrane, bacterial amylase, fungal amylase, cell wall, vinozyme, pectinase, polysaccharide/protein, denatured/coagulated protein micelles so that functional ingredients can be easily extracted out (Zang et al. 2018).

Agitation Extraction (AE)

Agitation extraction is performed by taking a known quantity of sample in 80% aqueous methanol in 1 : 4 ratio in Erlenmeyer flask and kept on a shaker at 200 rpm for 30 min at room temperature in order to release free and bound phenolic compound forms. After it centrifugation at 10000 ×g for 3 min is done and to take supernatant which is evaporated in a rotary evaporator at 37°C and the dried content is dissolved in 10 mL MeOH. This way the free phenolic compounds is obtained. After this extraction, alkaline hydrolysis is done by adding 20 mL of 3 M NaOH directly to the residue and agitated on shaker for 4 hrs followed by acidified to pH 2 with 4 M HCl. The clear solution is extracted with 10 mL of ethyl acetate and is evaporated to dryness under vacuum in a rotary evaporator at 37°C. The dried residue is dissolved in 10 mL MeOH (Zang et al. 2018).

High Hydrostatic Pressure Extraction (HHPE)

This technique involves a high hydrostatic pressure system (17 MPa s−1) for 5 s for extraction using water as medium in 2 L processing unit. The packaged samples (known quantity) in HDPE pouches are extracted with 80% methanol. The packaged samples were placed in a cylindrical loading container at room temperature and pressurized at given pressure and time. After it, the extraction of free and bound form of ingredients is collected (Zang et al. 2018).

High Hydrostatic Pressure-Agitation Extraction (HHPE-AE)

In this technique two extraction techniques i.e. high hydrostatic pressure and agitation are performed together. The packaged samples (known quantity) in HDPE pouches are extracted with 80% methanol. The packaged samples were placed in a cylindrical loading container at room temperature and pressurized at given pressure for a time period. Then, agitation is carried out on shaker at 200 rpm for 15 min. After it, the extraction of free and bound form of ingredients is collected (Zang et al. 2018).

High Hydrostatic Pressure-Ultrasound Extraction (HHPE-UE)

In this technique two extraction technique i.e. high hydrostatic pressure and an ultrasound exposure is used for extraction. The packaged samples (known quantity) in HDPE pouches are extracted with 80% methanol. The packaged samples were placed in a cylindrical loading container at room temperature and pressurized at given pressure for a time period. Then, again extraction is done by ultrasound for 15 min at 60 Hz (Zang et al. 2018).

Microwave Assisted Extraction (MAE) or Green Technology

In this method microwave energy is used at 720 W with intermittent cooling for certain time to extract out functional ingredient from samples. Microwave is electromagnetic radiation with a frequency from 0.3-300 GHz. Microwave heat disrupts weak hydrogen bonds due to dipole rotation of molecules and the migration of dissolved ions increases the penetration of solvent in to the sample in screw capped brown glass bottle and thus extraction takes place. This technique is also called as green technology because this technique does not use any organic solvent to extract out functional ingredient from samples (Wang and Bohn 2012).

Advantages:

- It needs less solvent consumption
- Fast operation time.
- High yield of product

Microwave Assisted Extraction (Continuous) (MAEC)

MAEC is done in the microwave oven at 720 W in screw capped bottle without any intermittent cooling. To minimize evaporation cap is kept little tighter than during MAE (Wang and Bohn 2012).

Ultrasonication Assisted Extraction (UAE)

This method is used in an ultrasonicator at 720 W in a glass beaker. In an ultrasonic cleaning bath, samples are taken with 70% ethanol for one hour at a frequency of 100 kHz at 25 ± 3°C to yield ultrasonic extract. By this way, due to disruption of cell walls, penetration of solvent occurs. The ultrasonic extract is centrifuged at 10,000 rpm for 15 min followed by filtration with Whatman no.1 filter paper is done. The filtrate is dried in a rotary evaporator. The dried extract is then taken in respective solvents and stored under refrigeration (4-8°C) (Wang and Bohn 2012).

Advantages

- Simple technology.
- Inexpensive method.
- Efficient than conventional microwave assisted method.
- Thermolabile constituents can be extracted out.
- Ultrasound apparatus is cheaper and easy in operation as well.

Purification of Extracted Functional Food Ingredients

There are various traditional and modern methods for the purification of extracted Functional Food Ingredients. These are described below (Fig. 7):

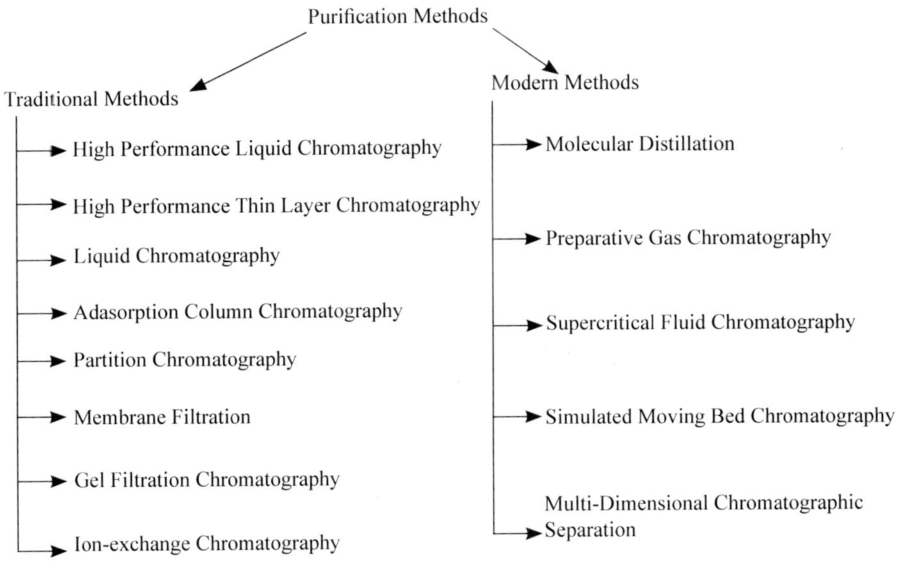

Fig. 7: Purification methods for Extracted Food Functional Ingredients

Traditional Methods

High Performance Liquid Chromatography (HPLC)

HPLC is an analytical technique which separates, identify and quantify the samples on the basis of the interaction with the solid particles of a tightly packed column and the solvent of the mobile. Hence, different constituents of a sample are eluted at different time period. High performance liquid chromatography is an improved version of column chromatography where solvent is forced through under high pressures of up to 400 atmospheres to move sample over column with solvent. Thus it is a faster method for qualitative and quantitative analysis (Fig. 8).

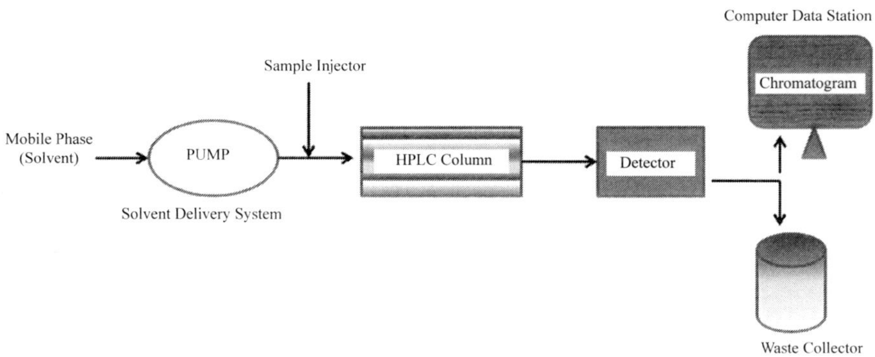

Fig. 8: Diagrammatic representation of High Performance Liquid Chromatography (HPLC) Instrumentation

High Performance Thin Layer Chromatography (HPTLC)

High-performance thin-layer chromatography is an improved version of thin-layer chromatography (TLC) which increases the resolution and allows more accurate quantitative measurements. It is also called Planer Chromatography or Flat Bed Chromatography. The principle of separation is adsorption. HPTLC is different from TLC in terms of particle size and pore size of the sorbents. This technique involves high quality TLC plates with finer particle sizes in the stationary phase to allow better separation. HPTLC functions on the principle of adsorption. In HPTLC, solvent flows with capillary action and separation of the components occurs based on their affinities towards the adsorbent (Coskun 2016).

Advantages:

- Fast separation method.
- Inexpensive technology.
- No need of time taking pretreatment steps.
- Development time is 40% less than TLC.
- The results are reproducible which is difficult in TLC.
- Number of samples can be taken more up to 50 per plate than TLC.

Liquid Chromatography (LC)

Liquid chromatography separates molecules in a low viscosity liquid mobile phase using a solid stationary phase made up of an immiscible liquid coated onto a porous support. In this technique, mass transfer of a sample takes place through a polar mobile phase and non-polar stationary phase (Coskun 2016).

Adsorption Column Chromatography

This technique performs adsorption of the solutes of sample over a adsorbent or stationary phase i.e. silica gel, resins etc. due to affinity and separates the mixture into individual components based on differences in adsorption affinity of sample components. Silica gel is most popular adsorbent. It adsorbs sample components hydrogen bonds and dipole–dipole interactions (Coskun 2016).

Partition Chromatography (PC)

Partition chromatography is liquid–liquid extraction method. In this technique, separation occurs on the basis of relative solubility of the liquid phase. Stationary phase is made up of by coating a liquid over a solid matrix such as silica gel, cellulose etc. and other liquid acts as a mobile phase. PC separates mixtures of closely related chemicals of the sample (Coskun 2016).

Membrane Filtration (MF)

In this technique, a semi-permeable membrane acts as specific filter which allows smaller molecules to pass through and retains the larger molecules using selective separation. Base on the filter mesh size, membrane filtration may be categorized as microfiltration, ultrafiltration, and nano-filtration. The sequential microfiltration, ultrafiltration, and nano-filtration can separates different sample components.

Gel Filtration Chromatography (GFC) or Gel Permeation Chromatography (GPC) or Size Exclusion Chromatography (SEC)

This technique is based on separation of sample solutes by the differential exclusion or inclusion of solutes while passing over stationary phase containing heteroporous cross-linked polymeric gel or beads such as Sephadex. The small molecules retain for longer time in GFC than large molecules (Coskun 2016).

Ion-exchange Chromatography (IEC)

Ion-exchange chromatography (IEC) separates molecules on the basis of the differences in net surface charges of sample components. Hence, the sample components having ionizable functional group can be separated out by this technique. The ionized components is absorbed and desorbed by ion-exchange resin by changing the ionic strength of the mobile phase by changing pH or salt concentration. Cation ion-exchange resins are used for anion separation and anion ion-exchange resins are used for cationic solutes separation (Coskun 2016).

Modern Methods

Molecular Distillation (MD)

It is a vacuum distillation process of separation, purification and concentration of thermo-sensitive sample components as per the distillation under high vacuum at a high temperature below than its boiling point for shorter time period. Thus, low or no decomposition of sample occurs.

Preparative Gas Chromatography (Prep-GC)

It is an analytical technique used to purify large quantities of sample in large-diameter packed columns over which a carrier gas flows continuously. This system consists of a carrier gas system, an injection system, a separation column splitter and detector (Fig.9) (Coskun 2016).

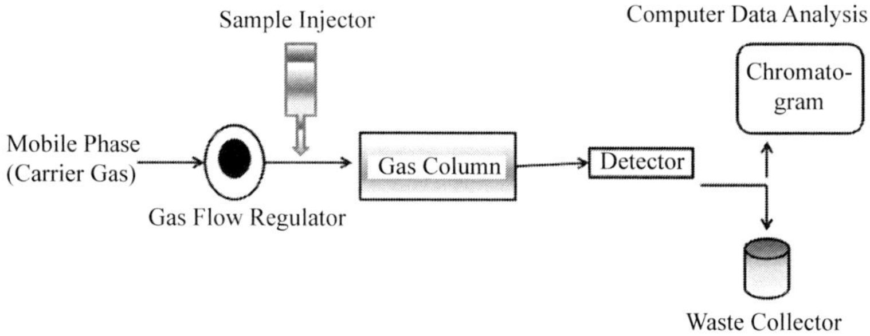

Fig. 9: Diagrammatic representation of Preparative Gas Chromatography (Prep-GC) Instrumentation

Supercritical Fluid Chromatography (SFC)

The instrumentation of this technique was manufactured by Hewlett-Packard in 1982. This technique uses supercritical fluid such as CO_2 and water as mobile phase for separation which allows rapid and efficient separation due to its high dissolved capability, high diffusivity, no surface tension and low viscosity. A supercritical fluid has both gas-and liquid-like properties. This chromatography has desirable characteristics of GC and HPLC. By this technology low polar and high molecular weight compound can be separated out.

Simulated Moving Bed Chromatography

Simulated moving bed (SMB) chromatography is also one of the separation technologies which use multiple columns with stationary phases (bed) for continuous multi-column chromatographic process. It is a powerful tool for the separation of sample components with less solvent consumption and shorter period of time. This technique is used for two component mixture in which chromatogram is obtained with leading and trailing key product peak.

Advantages

- It is a powerful tool for the separation of sample components
- It needs less solvent consumption
- It takes shorter period of time

Multi-Dimensional Chromatographic Separation

Multidimensional chromatography allows separation of complex mixtures by using already coupled multiple columns with different stationary phases which allow transfer of fractions from the first column to other columns for additional separation. This enables separation of complex mixtures that cannot be separated using a single column (Coskun 2016).

Effect of Food Processing Methods on Stability of Functional Food Ingredients

Functional Ingredients present in food undergo different types of inconvenient chemical changes during technological processing such as antioxidant oxidation, complex formation of food ingredients with other food compounds, enzymatic modifications and changes from active antioxidant into pro-oxidative form etc. Some of changes are undesired and cause a decrease of nutritional value of food products. Technological processing includes various thermal, hydrothermal, culinary processing methods (Jim and Edward 2010). The list of all food processing technologies is presented below (Fig. 10).

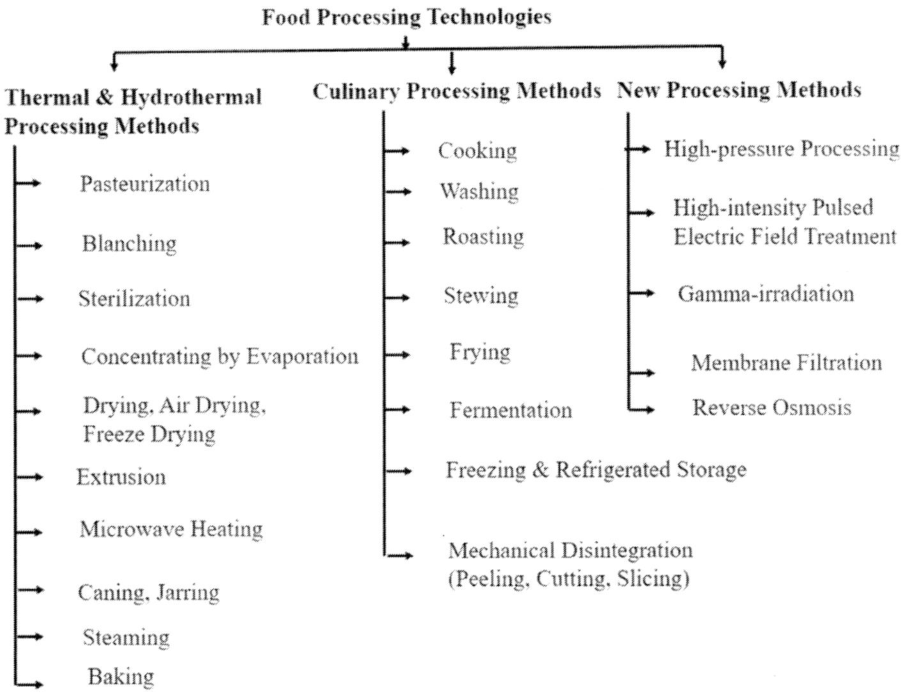

Fig. 10: Flow diagram of Food Processing Technologies

Thermal, Hydrothermal & Culinary Processing Methods

Thermal processing is the most common and traditional technique used for food processing. Thermal processing is defined as the combination of temperature and time required to eliminate a desired number of microorganisms from a food product. Thermal processing entails the application of heat energy for food transformation into the desired safe and stable forms.

Hydrothermal processing is mainly food preparation method at home. In this processing method, water environment enhances fast heat transfer to the whole product volume leading to equal heating of the product. This processing may lead to the decrease of antioxidant potential. Mild heating may enhance and high-temperature treatment may decrease the content of pro-oxidative compounds.

Following are some thermal and hydrothermal processing methods and their impact on the stability of functional food ingredients described:

Blanching

Blanching is a hydrothermal cooking process in which a food is kept in boiling water for a definite time period and followed by placed in cold water to halt the cooking process. Blanching causes structure looseness, loss of firmness, leaching of soluble compounds from the food surface, partial inactivation of oxidizing enzymes, losses of antioxidants as well.

Pasteurization

Pasteurization is a thermal process in which foods are treated with mild heat to eliminate pathogens and extend shelf life. It causes denaturation and precipitation of protein, inactivation of enzymes, reduced food firmness, thermal degradation and leaching of functional food ingredients.

Air Drying

Air drying is a process of decreasing the amount of steam in the air using appropriate methods. The humidity can be directly removed from the air by: Lowering the air temperature below the dew point, which makes the water condense. This method is used in air dryers with a refrigeration system. Air drying causes volume reduction, porosity reduction, denaturation of protein, contraction of surface layer, temperature increase, oxidation, thermal degradation, interaction of polyphenols with proteins and carbohydrates, formation of Maillard Reaction Products (MRP) with pro- and antioxidant activity, cis-isomerisation of carotenoids, dramatic decrease of vitamin C and tocopherols with increase of temperature over 80◦C and increase in -carotene content of food products (Jim and Edward 2010).

Extrusion

Extrusion is a kind of hydrothermal processing. It involves specific conditions of treatment, namely, high pressure combined with high temperature. Extrusion is performed at temperatures reaching 160–220°C, which destroys vitamins C, E and A. Therefore, this processing method is drastic with regards to impact on active compounds. It allows a reduction in losses of antioxidants and volatile aroma substances.

Microwave Heating

Microwave heating involves electromagnetic waves and heat transfer in the food products. It causes volume reduction, porosity reduction, inhibition of enzymes, protein denaturation, oxidation, thermal degradation, formation of MRP with pro and antioxidant activity, decrease of vitamin C and tocopherols with increase of temperature over 80°C, increase in β-carotene content of food items.

Canning and Jarring

Canning and jarring are the methods of storing and preserving food for the long time period. These methods cause structure looseness, protein denaturation, leaching of soluble compounds, inactivation of enzymes, thermal degradation of antioxidants and leaching of antioxidants from the foods.

Baking, Frying and Roasting

Roasting is cooking food on a low heat for a long amount of time, baking is usually done in an oven where food is baked at a high temperature for a short amount of time, and frying is cooking food at a high temperature in fats like oil, lard, etc. These processing techniques cause volume reduction, porosity reduction, saturation with fat, hydrolysis of glycerols, high resistance against destructively high temperature (180◦C), oxidation of lipids affecting flavour, taste, colour and aroma, thermal polymerization of secondary oxidation products, darkening polymer formation, thermal degradation, oxidation, dramatic losses in vitamin C, antioxidants loss and increase of flavonoids during baking.

Cooking

Cooking is the practice of preparing food with the use of heat or cold for consumption. It causes looseness in structure, firmness, tissue disintegration, denaturation of membrane protein, inhibition of enzymes, leaching of soluble compounds, softening of tissue of foods, antioxidant leaking, thermal degradation, release of some carotenoids, small reduction in total antioxidant capacity, significant loss of vitamin C etc.

Mechanical Disintegration (Peeling, Cutting, Slicing)

Peeling, cutting and slicing of food products causes mechanical damages to the tissue leading to the exposition of hypodermal tissue to oxygen and microorganisms. It also causes surface dehydration, moisture loss, high respiratory rate and thus loss of firmness of freshness of food products. Hence, the mechanical disintegration cause oxidative reactions for cellular biomolecules and also responsible for quantitative losses of functional food ingredients due to leakage problem.

Washing

Washing fresh fruits and vegetables is a proven way to remove germs and unwanted residues from their surfaces before eating them. It also causes hydration and lowering the product temperature. Properly washing fresh fruits and vegetables is a good habit to practice to minimize the ingestion of potentially harmful residues and germs. Health and food safety experts, including the Food and Drug Administration (FDA) and Centers for Disease Control (CDC), strongly recommend consumers to wash the foods with plain water only. Vigorous washing practice may lead to disruption of outer surface of foods and hence it may cause leakage and oxidative reaction of functional food ingredients.

Freeze Drying

Freeze drying or lyophilization or cryodesiccation, is a low temperature dehydration process which performs freezing food product, lowering pressure and removing the ice by sublimation process. This process causes damages in plant tissue by ice crystals formation, oxidation, and evaporation of functional food ingredients of food products (Jim and Edward 2010).

Freezing and Refrigerated Storage

Refrigeration is a non-thermal process which slows down the chemical and biological processes in foods. It also reduces deterioration and the loss of quality. The storage life of fresh perishable foods such as meats, fish, fruits, and vegetables can be extended by several days by cooling, and by several weeks or months by freezing. This process causes damages in plant tissue by ice crystals formation, oxidation, leaching after thawing and precipitation of flavanones of food products.

Fermentation

Fermentation is a biochemical.metabolic process.by which molecules such as glucose are broken down anaerobically through the action of enzymes. Microorganisms like yeast and bacteria break down.food.components (e.g. sugars such as glucose) into other products (e.g. organic acids, gases or alcohol) with unique and desirable taste, aroma, texture and appearance of foods. Fermentation causes increase of moisture, structure looseness, antioxidant losses, glycoside hydrolysis, enzymatic oxidation and enzymatic degradation of antioxidants of foods.

Non-thermal Methods

An adverse effect on organoleptic properties of foods caused by thermal processing. Therefore, milder processing techniques including novel thermal, minimal thermal and non-thermal processing has become the new trend. These techniques provide food of high quality and with improved functionality and produce safe and better quality food products.

High Pressure Treatment or Pascalization

High pressure processing (HPP) or Pascalization is a method of preserving and sterilizing food in which a product is processed under very high pressure (100–1000 MPa) for short periods of time (20 min) leading to the inactivation of certain microorganisms and enzymes in the food. High Pressures applied in food products causes disruption of tissue structure of matrix and cell membranes leading to release of antioxidants from the food matrix, homogenization, denaturation of protein, inactivation of enzymes, slight losses due to partial degradation and depolymerisation, improvement of extractability of carotenoids of food items (Jim and Edward 2010).

Pulsed Electric Field (PEF) Technology

It is a non-thermal food preservation method that involves the use of short electricity pulses for microbial inactivation while imposing minimal detrimental influence on food quality. This treatment involves applying very short (10–1000 s) electric pulses at high electric intensities (20–80 kV/cm). This technology causes concentration of dry matter, volume reduction and small thermal degradation of antioxidants content of foods. This method also inactivates peroxidase, which is involved in the oxidation of a wide range of natural compounds. General assessment of the technique is positive because it destroys antioxidants at a very low level (Jim and Edward 2010).

Irradiation

Food irradiation is the application of ionizing radiation to food which improves the safety and extends the shelf life of foods by reducing or eliminating microorganisms that cause food poisoning, such as *Salmonella, Campylobacter* and *E. Coli*. Allowed doses of radiation is 30 kGy (http://www.iaea.org/icgfi). The range of ionizing radiation usage is increasing, especially because of the ban on fumigation with ethylene oxide. According to widely accepted data, irradiation does not make food products radioactive and radiolytic products formed during irradiation pose no danger to humans. Irradiation causes the increase of cell permeability, antioxidant alterations, oxidation and partial degradation of antioxidants of foods (Jim and Edward 2010).

Microencapsulation

Microencapsulation is a process in which tiny particles or droplets are surrounded by a coating to give small capsules with useful properties. Microencapsulation reduces the reactivity of the active material in the external environment and speed of losses and evaporation of the core material into the medium. It improves food handling, provide controlled release of the active product and mask unpleasant odor. Microencapsulation protects.antioxidants. from heat, light and oxygen, allowing stability, bioavailability, flavour masking and controlled release while maintaining their functional properties and increasing ease of handling (Jim and Edward 2010).

Conclusion

Nowadays, consumers are very much aware about the components contained in the foods they are consuming. Hence, they prefer those foods which are contained with all necessary functional food ingredients which provide health benefits. Thus, functional foods are gaining importance by consumers due to changes in eating habits and concern about health. In this regard, the present chapter has been designed which focuses on various approaches of processing (preparation and purification) of functional food ingredients from various natural sources such as cereals, oilseeds, legumes, vegetables, fruits, herbs, spices fish, poultry, meat, egg, milk, microorganisms etc. Besides the topic covered in the chapter, more focus has to be needed to give on some points such as packaging, preservation and storage of functional food ingredients. Moreover, more research is needed in the near future to demonstrate the in vivo stability of extracted and purified functional food ingredients has to be discussed upon incorporating them in food in order to design of new functional food. This will help World Health Organization regulations for approval and greater use of these food items.

The oxidative stability of foods is a very complicated matter because it demands the control of pro-oxidative and antioxidant compounds. It is often difficult to choose the best processing technique. The use of thermal processing leads to pro-oxidative enzyme inactivation,

Among methods that are less harmful to antioxidants are new physical methods such as high-pressure and pulsed electric field processing, as well as low-temperature vacuum drying, puffing, short cooking in small water amounts and fermentation processes

To assess the protective role of antioxidants in decreasing the risk of cancer and cardiovascular disease, there is a fundamental need to study the role of interactions between chemical changes and biological activity of antioxidants in human beings. Expanding knowledge in this area should lead to improvements in technological processing methods in such a way as to produce high-quality food with effective biological activity. In this respect, genetic engineering has much to offer because it could allow the engineering of plant materials that would supply high antioxidant capacity, while reducing losses due to oxidation. These methods could also lead to higher functional activity of food products. There is also a widening field in enzymatic processing of food matrices in the direction of increasing antioxidant extractability, and also structure modification to increase natural product diversity and produce more active and easily available antioxidants (Willits et al. 2004). In the growing long-term consciousness of consumers, technological development should lead to an increase in the role of the functional food segment.

References

Alvarez-Rivera G, Bueno M, Vivas DB et al. 2020. Pressurized Liquid Extraction. In book: Liquid-Phase Extraction

Azwanlda NN. 2015. A Review on the Extraction Methods Use in Medicinal Plants, Principle, Strength and Limitation. Medicinal and Aromatic Plants: 4, 3.

Carullo D, Pataro G, Donsi F, Ferrari G. 2020. Pulsed Electric Fields-Assisted Extraction of Valuable Compounds from *Arthrospira Platensis*: Effect of Pulse Polarity and Mild Heating. Front. Bioeng. Biotechnol.

Coskun O. 2016. Separation techniques: Chromatography. North Clin Istanb. 2016; 3 (2):156-160.

Gusthinnadura Oshadie De Silva, Achala Theekshana Abeysundara, Malamige Minoli Weroshana Aponso. Extraction methods, qualitative and quantitative techniques for screening of phytochemicals from plants. Am J Essent Oil Nat Prod 2017; 5(2):29-32.

Jim Smith and Edward Charter. 2010. Functional Food Product Development, pp 1-536 Blackwell Publishing Ltd ISBN 978-1-4051-7876-1.

Mital J. Kaneria & Manisha B. Bapodara & Sumitra V. Chanda, 2012. Effect of Extraction Techniques and Solvents on Antioxidant Activity of Pomegranate (*Punica granatum* L.) Leaf and Stem. Food Anal Methods, 5:396-404.

Mottaleb M.A., Abdul M. 2012. Accelerated Solvent Extraction for Natural Products Isolation. Methods in molecular biology (Clifton, N.J.) 864:75-87DOI: 10.1007/978-1-61779-624-1_4

Sánchez-Camargo A.P. et al. 2014. Supercritical Fluid Extraction. In Reference Module in Chemistry, Molecular Sciences and Chemical Engineering.

Wang L, Bohn T. 2012. Health-Promoting Food Ingredients and Functional Food Processing. In book: Nutrition, Well-Being and Health, DOI: 10.5772/25862

Zhang, QW, Lin, LG. & Ye, WC. Techniques for extraction and isolation of natural products: a comprehensive review. Chin Med 13, 20 (2018).

5

Natural Bioactive Compounds in Fruits

[1]Om Singh, [2]P. Koli, [3]B. H. Choudhury and [4]Richa Singh

[1]Department of Post-Harvest Management
R.V.S.K.V.V. Campus, College of Horticulture, Mandsaur, Madhya Pradesh
[2]PAR Division, ICAR-IGFRI, Jhansi, Uttar Pradesh
[3]Department of BAC, B. N. College of Agriculture, BNCA, AAU
Jorhat, Assam
[4]Krishi Vigyan Kendra, Amberpur, Sitapur, Uttar Pradesh

Introduction

Fruits contain various beneficial compounds that possess biological activities and can be used as pharmaceutical nutraceutical products. Most of the fruits contain polyphenolic components that can promote antioxidant potential on human health. Wild Blueberries, strawberries, plums, oranges, grapes, cherries and mangoes are rich in antioxidant. Additionally, anti-inflammatory, antibacterial, antifungal and chemopreventive effects are associated with the various fruits. Besides polyphenols, other compounds such as xanthones, carotenoids, saponins also exhibit health effects. Fruit consumption is considerably increasing in the daily diet because they supply high levels of biologically active compounds that impart health benefits beyond basic nutrition. Fruits provide an optimal mixture of antioxidants such as vitamins C and E, polyphenols and carotenoids, vitamins, phytoestrogens, glucosinolates and anthocyanins. Diets rich in fruits may help to delay the ageing processes and reduce the risk of certain types of cancer, cardiovascular diseases and other chronic diseases. The combination of vitamins, minerals, phenolic, antioxidants and fibre seems to be responsible for these effects. The most dietary components in fruits acting as antioxidants are fibre, polyphenols, conjugated isomers of linoleic acid, D-limolene, epigallocatechin, gallate, soy protein, isoflavones, vitamins A, B, C and E (group of tocopherols), calcium, selenium, chlorophyllin, alipharin, sulphides, catechin, tetrahy-drocurecumin, seasaminol, glutathione, uric acid, indoles, thiocyanates and protease inhibitors.

Antioxidant substances protect other compounds from oxidation by being oxidized themselves. Antioxidants act as preservatives that prevent the fats in foods from becoming rancid and prevent other damage to food caused by oxygen. In the body, smoking, sunlight and pollution increase oxidation. Antioxidants protect against the free radical molecules that are produced by the breakdown of food in the body and by environmental exposure to tobacco smoke or radiation. Antioxidants have different effects when consumed by humans; they help to control blood pressure or blood sugar-influencing substances, or can act as agents with anti-carcinogenic, immunity supporting, antibacterial, antifungal, antiviral, cholesterol-lowering, antithrombotic or anti-inflammatory properties.

There is a lack of information regarding the content of bioactive compounds of other important fruits, such as exotic and tropical fruits. Although the health benefits that promote their consumption are well known, information on their bioactive compounds is required to know their specific contribution to the total antioxidant capacity of the tissues. Fruits like banana, mango, pineapple, papaya and guava are the most popular tropical fruits all over the world due to their bright colour, characteristic taste and nutritional value. These fruits are exported to the USA and Europe from subtropical and tropical countries in response to their increasing demand. The purpose of this article is to update the information regarding the bioactive compounds and antioxidant capacity of fruits, as well as to disseminate knowledge on the health benefits of the most commonly consumed fruits in the international markets, such as apples, banana, mango, pineapple, papaya, berries and guava.

Bioactive compounds

A type of chemical found in small amounts in plants and certain foods (such as fruits, vegetables, nuts, oils, and whole grains). Bioactive compounds have actions in the body that may promote good health. They are being studied in the prevention of cancer, heart disease and other diseases. Examples of bioactive compounds include lycopene, resveratrol, lignan, tannins, and indoles. The term "nutraceutical" combines with the word "nutrient" (a nourishing food or food component) with "pharmaceutical effect" (a medical drug) in 1989 by Stephen De Felice, (MD, Founder and Chairman of the Foundation for Innovation in Medicine). According to De Felice nutraceutical can be defined as a food (or part of a food) that provides medical or health benefits including the prevention or treatment of a disease. Nutraceuticals may contain substances that are "natural" expressed intent of treatment or prevention of disease but may not be generally recognized as safe (Ross 2000).

Table 1: Bioactive compounds and their nutraceuticals properties classified as per natural sources

Components	Fruit source	Potential benefit
Carotene		
Lycopene	Pink Grapefruit, Guava, Papaya, Watermelon	Antioxidant activity, protects against formation of cancer mainly prostate, bladder, cervical, leukemia
β-carotene	Apricot, Oranges, Mango, Papaya	Pro-vitamin A activity, Anti-oxidant prevent cancer and check conversion of LDL cholesterol into its atherogenic form
α-carotene	Carrot, Lemon, Water melon, Papaya, Banana	Antioxidant, Anti cancer potential.
Zeoxanthine	Avocado, citrus	Protects eyes against muscular degeneration
Polyphenols		
Anthocyanidine	Fruit berries	Neutralizes free radicals, reduce risk of cancer, counteracts inflammation in the body, lower blood sugar
Flavonones	Citrus	Antioxidant, Anti cancer activity
Flavones	Fruit berries	Antioxidant, Anti cancer activity
Flavonols	Apple	Antioxidant activity
Phenolic acids	Berries	Antioxidant, Anti cancer activity
Resveratrol	Dark grapes, Raisins, Berries	Lower total sugar, Serum cholesterol
Phenols		
Ferulic acid	Apple, Pears, Citrus fruits	Healthy vision and heart

Bioactive compounds in berries

Berries are a group of small fruits that grow in cold weather regions such as North America and Europe and some high altitude locations of southern countries. They are characterized by their small size and dark colour. Common berries include fruits of the genus *Vaccinium* spp (blueberry), *Fragaria* (strawberry), *Rubus* (raspberry, black berry), *Ribes* (red currant, black currant, goose berry), *Cornus* (cornelian cherry), *Prunus* (sour cherry), *Aronia* (chokeberry), *Sambucus* (elderberry). Berries are rich in a wide range bioactive compound such as flavonoids (flavan-3-ols, flavonols, anthocyanins, procyanidins) and phenolic acid and its derivatives (hydroxycinnamic, ellagic) that possess antioxidant activity. As mentioned above, anthocyanins and other flavonols are the main phytochemicals found in berries (Ferlemi and Lamari 2016).

Phytochemicals consisted of two words *Phyto* and *Chemicals*, *Phyto* from Greek word for plants, denoting their plant origins and it refers to the chemicals which are present in plants. These phytochemicals are secreted or developed as a part of the plants own defense mechanism against environmental insult and fortunately beneficial to human beings. These phytochemicals are very effective antioxidants thus checks free radical mediated degenerative diseases like cancer. Health professionals are gradually recognizing the role of phytochemicals in health improvement. The beneficial health effects of fruit phytochemicals are associated with the neutralization of these reactive oxygen molecules.

Bioactive compounds in tropical fruits

Table 2: Different Colored fruits provide different Phytochemicals

Color Group	Phytochemicals/Potential Health Benefit	Fruits
Red	Anthocyanin – delay several diseases associated with ageing	Watermelon, Strawberries, Papaya and Pink Guava
Yellow/Orange	Carotene and flavonoids – reduce risk of heart disease, cancer and slow down ageing	Mango, Oranges
Blue/Purple	Anthocyanins and phenolics – reduce risk of cancers, heart disease and slow down ageing	Grapes, Plums, Blueberries
White	Polyphenols –reduce risk of heart disease	Bananas

When consumed regularly, these fruits can be a valuable dietary source of many phytochemical compounds, which also impart a characteristic colour and flavour to the fruit. The most common bioactive compounds present in tropical fruits are vitamin C and E, carotenoids, flavonoids and thiol compounds (Gonzalez-Aguilar et al. 2008).

Table 3: Total phenolics and antioxidant activity of the most commonly consumed fruits

Fruit	Total phenolic content (mg /100gram fresh fruit)	Total anti-oxidant activity (μmol/gram)
Cranberry	527.2	177.0
Apple	296.3	97.6
Red Grape	201.0	64.7
Strawberry	160.0	64.4
Peach	84.6	49.5
Lemon	81.9	42.8
Orange	81.2	31.5
Pear	70.6	34.2
Grape fruit	49.6	24.7

Among these, polyphenolics are widely distributed secondary metabolites that serve as the predominant antioxidant compounds. Several studies have reported polyphenolic compounds in mango, papaya and pineapple flesh and peel, including various flavonoids, xanthones, phenolic acids and gallotannins. Among these compounds, mirecitine, mangiferine, gallic acid and hydrolysable tannins, which are most likely gallotannins, constitute the major antioxidant polyphenolics found. The amount of bioactive compounds present in tropical fruits varies widely. Banana has the lowest vitamin C, E and β-carotene levels compared with mango, papaya, and pineapple. However, banana has higher phenol content than pineapple and papaya. It appears that the high phenol content present in mango and banana contributes to a higher extent to their total antioxidant capacity than other bioactive compounds (Gonzalez-Aguilar et al. 2008).

Table 4: Amount of bioactive compounds present in fresh tropical fruits

Fruit	Vitamin C (mg/100 g)	β-Carotene (mg/100 g)	Vitamin E (mg/100 g)	Total phenols (mg GAE*/g)
Banana	8.9	26.0	0.1	2.3
Mango	27.0	445.0	1.1	2.6
Pineapple	36.0	34.0	0.07	1.7
Papaya	61.8	276.0	0.7	0.28

*GAE: Gallic Acid Equivalents

Bioactive compounds in apples

Apples are one of the most frequently consumed fruits in the world, constituting an important part of the human diet as a source of sugars, minerals and dietary fibres. Apples are an important source of biologically active substances, such as vitamin C, flavonoids and phenol compounds, all of which are known to act as natural antioxidants. It is for this reason that apples are considered one of the major contributors of phytochemicals in the human diet.

Table 5: Average concentrations of major bioactive compounds in apple* and their contribution to the total antioxidant activity

Bioactive Compounds	Concentration (mg/100g fresh weight of apples)	% Relative contribution
Quercetin glycosides	13.2	34.7
Procyanidin B2	9.35	19.0
Chlorogenic acid	9.02	7.6
Epicatechin	8.65	19.9
Phloretin glycosides	5.59	7.8
Vitamin C	12.8	11.0

*Average of six apple cultivars: Rhode Island Greening, Golden Delicious, Cortland, Monroe, Empire and NY674.

These compounds were also associated with oxidative deterioration, responsible for the discoloration of apple and apple products. Phytochemicals, bioactive compounds found in apples are suggested to possess health benefits such as antioxidant and antiproliferative activities. Overproduction of oxidants in humans can cause oxidative stress, which is associated with chronic diseases; therefore, increased consumption of fruits containing high levels of antioxidants could prevent or slow the oxidative stress caused by free radicals. Phytochemicals in fruits can have complementary and overlapping mechanisms of action, acting as antioxidant agents, stimulants of the immune system, regulators of gene expression in cell proliferation and apoptosis, intermediaries in hormone metabolism, as well as antibacterial and antiviral agents. Apples have the ability to decrease the incidence of some chronic diseases due to the complex mixture of phytochemicals present in the fruits with additive and synergistic effects that enhance their antioxidant and antiproliferative activities. Bioactive compounds contained in apples include: carotenoids, flavonoids, isoflavonoids, phenolic acids, lignans and triterpenes (Dolkar et al. 2017).

Conclusion

It is well documented that regular consumption of fruits, including berries, apples and tropical fruits, as part of a healthy diet may aid in the prevention of chronic diseases and foster good health. In general, berries are a major source of antioxidant compounds, followed by apples and tropical fruits. Berries are especially rich in phytochemicals and anthocyanins, and are therefore a main source of these health promoting substances. There is an increased global interest due to the recognition that nutraceuticals an important role in health enhancement and there is great awareness among the people for the beneficial effect of nutraceuticals in day to day life. Fruits contain significant amounts of bioactive components as they provide desirable health benefits beyond basic nutrition and play vital role in the prevention of diseases.

Reference

Dolkar, D., Bakshi, P., Wali, V.K., Sharma, V. and Shah, R. A. 2017. Fruits as nutraceuticals. Eco. Env. & Cons. 23: 113-118.

Ferlemi, A.-V.; Lamari, F.N. 2016. Berry Leaves: An Alternative Source of Bioactive Natural Products of Nutritional and Medicinal Value. Antioxidants 5, 17.

Gonzalez-Aguilar, G., Sanchez, RM R., Tellez, MA M., Olivas, GI, Parrilla, E A. and Rosa, LA de la. 2008. Bioactive compounds in fruits: health benefits and effect of storage conditions. Stewart Postharvest Review (An international journal for reviews in postharvest biology and technology). 3:8.

Ross, S. 2000. Functional foods: the Food and Drug Admin-istration perspective. American Journal of Clinical Nutrition. 71: 1735-1738.

6

Natural Bioactive Compounds in Vegetables

[1]Prachi Gera and [1]Priti Khemariya

Indian Institute of Packaging, Patparganj, Delhi-110092

Introduction

Functional elements are non-conventional biomolecules in foods that have the ability to modulate one or more body metabolic processes or pathways, leading in health advantages and well-being promotion. Research has demonstrated a connection between food, health and well-being functional elements. Consequently, functional elements have health-promoting roles connected with multiple progressive steps at various phases of disease control, from initiation to growth. They can therefore be implemented efficiently in disease therapy and prevention. Therefore, the borderline between food and medicine is becoming very thin at a moment when the role of a healthy diet in stopping non-communicable diseases is well recognized. Functional elements include plant-derived, non-nutritious and biologically active chemicals that operate in the body to avoid certain non-communicable diseases from occurring.

Food contains over 900 phytochemicals. One fruit or vegetable serving (about 120g) can contain as many as 100 distinct phytochemicals. Previously, functional elements were believed to happen predominantly only as phytochemicals in plant products including whole grains, fruits and vegetables. However, in animal goods such as milk, fermented dairy products and cold water fish, probiotics, conjugated linolenic acid, long-chain omega-3, −6 and −9 polyunsaturated fatty acids and bioactive peptides are also discovered. Usually, functional elements happen in various forms in food such as glycosylated, esterified, thiolated or hydroxylated materials. They also have numerous metabolic operations that enable positive impacts in the body on several illnesses and target tissues (Balic et al. 2020).

Functional Components Commonly Found in Food

Non - Starchy Carbohydrates

In essence, these are structural and storage carbohydrates that are polymers of glucose molecules and other sugars, including galactose, fructose, xylose, arabinose, etc., but are not starchy in nature (i.e. their sugar units are not connected by either α [1, 4] or α [1, 6]). Thus, human digestive enzymes do not hydrolyze them, but they are fermented by the probiotic microbes in the colon. A few non-starchy carbohydrates are.dietary fiber and fucoidan.

Dietary Fibres

Non-starchy polysaccharides and structural elements of the cell walls of cereals and microorganisms are dietary fibers (DF), which may be soluble or insoluble. They are the indigestible part of plant foods made up of long straight and branched chains of carbohydrate molecules held together by bonds that human digestive enzymes cannot hydrolyze. Chemically, DF are β (1, 3:1, 4) or β (1, 3:1, 6) hetero structural glucose polymers. The water-soluble fibers consist primarily of β-glucans, gums, pectin, mucilage and arabinoxylans, while the water-insoluble fibers consist of lignin, cellulose and hemicellulose.

Fucoidan

Fucoidan is a polysaccharide that is non-starchy and sulfated. It is a polymer of subunits associated with α (13) fucosepyranose sugar. There are also traces of galactose, xylose and glucoronic acid in Fucoidan. The absence of a hydroxyl group on the six-position carbon (C-6) and its L-configuration are two structural characteristics that distinguish fucose from other six-carbon sugars current in mammals.

The lengthy fibrous structures of dietary fiber enable them to trap into the digestive tract damaging toxins and carcinogens. The soluble nutritional fibre, cereal β-glucan, has received unique attention for its many health advantages such as reducing serum cholesterol. It is also ascribed to excellent water retention capacity, gelling capacity, and hydro-colloidal forming characteristics that have affected their use as fat replacements. Fungi β-glucans, a family of diverse structures contained in the yeast and mold cell wall, modulate the immune system by improving the activity of leucocytes responsible for improving the mechanism of body defense. Inulin substituted fat in milk goods effectively. Soluble nutritional fibers can dissolve or absorb water and are efficient in binding the intestinal tract toxins and cholesterol. On the other side, insoluble nutritional fibers are unable to dissolve in water and are efficient in adding faecal bulk and increasing the frequency of food passage through the intestinal tract. Insoluble nutritional fibers also dilute prospective carcinogens and decrease contact with the intestinal tract of toxins and carcinogens and accelerate their passage out of the body (Khoury et al. 2012).

Similarly, fucoidan inhibits the spread of cancer cells by stopping tumor cells from adhering to the extracellular matrix and by inducing apoptosis or programmed self-destruction in the type I (HTLV-1) human T-cell leukemia virus that causes adult T-cell cancer. By inactivating NF-kB, a naturally occurring substance which controls antiapoptotic proteins, the polysaccharide paves the way for apoptosis. It has also been shown that Fucoidan stimulates macrophage phagocytic action and synthesis of several kinds of immune cells, which increases protection against infection. Fucoidan's nutritional makeup could be compared to breast milk, which is recognized as the most ideal immune-supporting food. By improving phagocytosis, the method through which white blood cells attack and kill pathogens, the polysaccharide provides the immune system a large boost. Also, Fucoidan raises the amount of mature white blood cells circulating in the body, thus strengthening the first line of defense against infections and illnesses.

Foods wealthy in soluble nutritional fibers include apples, cranberries, mangoes, oranges, asparagus, broccoli, carrots, peanuts, walnuts, most legumes, oats, and psyllium, while those wealthy in insoluble nutritional fibers are apples, bananas, berries, broccoli, green peppers, spinach, almonds, sesame seeds, most legumes, brown rice, whole-wheat breads, and cereals. Cereals and fungi are wealthy sources of β-(1, 3:1, 4)-glucan (a DF with powerful colloidal characteristics deemed to be a useful functional component for cholesterol lowering and low-glycaemic index features in foods). The largest amount of β-(1, 3:1, 4)-glucan occurs in bran, aleurone and subaleurone layers of oat (2, 2–7, 8%) and barley (2, 5–11, 3%). Brown seaweeds are high in fucoidan and some medicinal mushrooms. Furthermore, for their antitumor activity, mushroom and yeast polysaccharides such as β-(1, 3:1, 6)-glucan have been focused and the chemical variety of these glycans ranges from homopolymers to extremely complicated heteropolymers. The formation of such polysaccharides involves varieties of sugars such as glucose, galactose, mannose, xylose, arabinose, sucrose ribose, glucouronic acid, etc. Some of the glycans form conjugates of greater powerful antitumor activity with proteins and peptides (Abuajah et al. 2015).

Antioxidants, anti-cancerous and immune-modulating agents

Antioxidants are groups of compounds in the cell that neutralize free radicals and species of reactive oxygen (ROS). A free radical is a carbon or oxygen atom with a heavily loaded and unstable unpaired electron. In lipids, proteins, and carbohydrates, free radicals can form. The following are examples of antioxidants:

Carotenoids

The carotenoids (e.g., lycopene, lutein) are lipid-soluble plant pigments that are either oxygenated or non-oxygenated hydrocarbons with at least 40 carbons and an extensive dual bond system. The predominant non-polar functional carotenoids are alpha-carotene, beta-carotene and lycopene, and lutein is the primary polar functional carotenoid. Carotenoids may be discovered in plant tissues to be esterified to fatty acids or un-esterified. With prospective chemo-preventive operations, lycopene is the most active oxygen neutralizer. The complete fruit and vegetable carotenoid content differs with era and storage (Khoo et al. 2011).

Polyphenols

The most numerous and widespread group of functional molecules are polyphenols. Polyphenols are various groups of plant substances containing one or more benzene rings and different groups of hydroxyl (OH), carbonyl (CO) and carboxylic acid (COOH).They frequently occur with one or more attached sugar residues in conjugated forms. Flavonoids are the most prevalent class of polyphenols. Other polyphenol kinds include catechins, arubingens, aflavins, isoflavones, and more than 8,000 others. The content of polyphenol can differ greatly from food source to food of the same sort. Some food products and fruits recorded the following ranges for complete polyphenol content: barley and millet (590–1, 500 mg/100 g dry matter); oats and maize (8.7–30.9 mg/100 g dry matter); Fresh onions and leeks (20–20, 25 mg/100 g dry matter), fresh brussels (6–15 mg/100 g dry matter), blueberries, strawberries, cranberries and raspberries with a total polyphenol content of approximately 37–429 mg/100 g dry matter (Cory et al. 2018).

Phytosterols

Phytosterols in humans are the plant cholesterol equivalent. They have comparable constructions. However, there are extra double bonds and methyl and/or ethyl groups in the side chains of plant sterols. Beta-sitosterol, campesterol, and stigmasterol are the most prevalent bioactive phytosterols. A non-vegetarian daily diet includes about 250 mg of unsaturated phytosterols, while a vegetarian diet includes more than 500 mg. Plant sterols saturated derivatives are plant stanols like sitostanol (Saheli et al. 2021).

Tocopherols and Tocotrienols

Tocopherols and tocotrienols are functional elements that are lipid-soluble and comprise a phenolic-chromanol ring connected to a side chain of isoprenoids that is either saturated (tocopherols) or unsaturated (tocotrienols). There are also four main types of tocopherols and tocotrienols, alpha, beta, gamma, and delta, which vary on the phenolic-chromanol ring in the amount and position of methyl groups. The tocopherols also have three asymmetric carbons at two, four, and eight positions of the side chain of the isoprenoid. There are therefore eight isomeric types of tocopherols, the most bioactive of which is RRR-a tocopherol and the most abundant in human blood and tissues (Rizvi et al. 2014).

Organo-sulphur Compounds

Cruciferous vegetables such as broccoli, cauliflower and brussel sprouts or allium plants (vegetables of the same class with onions and garlic) such as leeks are frequently discovered in organosulfur compounds. Organosulfur compounds comprise sulfur atoms in a cyclic or non-cyclic configuration that are linked to a group of cyanates or carbon atoms. The functional components of foods containing organosulfur compounds are only acquired after the cells have been interrupted by cutting, chewing or crushing to expose them. Different isothiocyanates such as sulforaphane, phenethyl isothiocyanate and benzyl isothiocyanate are created from glucosinolyates by myrosinase action in cruciferous vegetables. Allicin is created in alliums from alliin and then quickly transformed by allinase action into diallyl sulfide, diallyl disulfide or diallyltrisulfide. These hydrolytic breakdown products are the functional elements that promote health in both cruciferous and allium vegetables.

The main functions of antioxidants include regulating a cell's redox potential and reducing potential cell death and carcinogenesis initiators. Antioxidants are therefore anti-carcinogenic. The capacity for redox relates to the equilibrium of the cell's reduction and oxidization responses. Redox modifications within a cell can cause different molecular reactions such as apoptosis induction (cell death) and signal transduction activation (cell-to-cell transmission of messages). Therefore, in optimizing health and illness prevention, redox regulation of physiological and pathological procedures is essential. Other functional antioxidant compounds, such as the binding of N-nitroso compounds by polyphenols in tea, can attach to toxins or carcinogens in the intestinal tract, thus preventing their conversion or even absorption. Phytosterol / stanols lipid-lowering mechanism happens in the intestinal tract by sequestering cholesterol and decreasing its absorption. Epidemiological and experimental studies indicate that nutritional phytosterols can provide protection from most prevalent cancers in Western societies such as cancers of the colon, breast and prostate. The potential mechanisms that phytosterols give this protection include their impacts on the composition of the membrane, tumor and host cell tissues, signal transduction pathways that control tumor development and apoptosis, host immune functions, and host metabolism of cholesterol. The structural resemblance between several isoflavone metabolites and those of estrogens and estradiols indicates the potential for isoflavones to engage in estrogen-like biological activity. However, isoflavones or phytoestrogens have antagonistic oestrogen activity that leads to lower general exposure to oestrogen in premenopausal females and reduces the risk of breast cancer. Phytoestrogen-rich diets in postmenopausal females decrease the increase in hormone-sensitive concentrations of plasma cholesterol and bone loss. Similarly, the induction of enzyme structures detoxifying toxic chemicals including the detoxifying enzymes stage I (e.g. cytochrome P450 group of oxidases) and phase II (e.g. N-acetyl transferase, glutathione S-transferase, UDP-glucoronyl transferase, etc.) is believed to decrease one's susceptibility to mutagenic effects. Phase II detoxifying enzymes can be activated by functional food parts with antioxidant features through the antioxidant response element pathway such as the atomic factor [erythroid-derived 2]-like two also known as the Nrf2 pathway. Organo-sulfur compounds such as isothiocyanates, in specific sulforaphane, are powerful monoinducers of detoxifying enzymes in stage II, while diallyl sulfides from garlic preparations are inducers of detoxifying enzymes in both stages I and II. The primary immune-modulation mechanism is the various antioxidant capacities of polyphenols, tocopherols, carotenoids, isothiocyanates, and allyl sulfides, with lycopene being the most active oxygen neutralizer with potential chemical-preventive operations. These compounds together can decrease the deleterious impacts of reactive oxygen species (ROS) and free radicals that cause immune cells to die prematurely. In reducing complete cholesterol concentrations, garlic is discovered to be a superior phytochemical and is also reported to regulate arterial stiffness by enhancing excellent cholesterol: high density lipoprotein (HDL) and reducing poor cholesterol: low density lipoprotein (LDL). It also inhibits inducible nitric oxide synthase by decreasing protein and mRNA and thus encourages vasodilatation of blood vessels. Garlic has powerful immunopotential ability and by delaying the hypersensitivity reaction increases the natural killer activity and proliferation of T- lymphocytes. Aged garlic extract is discovered to be a promising immune modifier with inner body regulatory function, especially in controlling sarcoma-180 and lung carcinoma and platelet aggregation inhibition. Both the oil- and water-soluble garlic extract elements have demonstrated health advantages. Specifically, its oil extract decreases severe mental disorder and avoids blood coagulation even in diabetics while its water extract is efficient in the HG2 hepatoma cells ' cell cycle and viability. Given the inverted oxidant reactions of garlic extract, it seems likely to protect tissues from oxidative damage (Abuajah et al. 2015).

Food sources of polyphenols and flavonoids include vegetables, fruits, cereals, legumes, nuts, tea, wines and other fruit, vegetable and grain-based drinks. Carrots, squash, sweet potatoes and spinach are abundant in both beta-and alpha-carotene and dark green leafy vegetables such as kale, spinach, mustard and green beans. Lycopene is mainly discovered in tomatoes. Kaempferol or chlorogenic acid, which have antimutagenic operations, are other bioactive elements in tomatoes. This indicates that the suspension of tomatoes has a protective impact on colon cancer that is mediated during carcinogenesis by modulating various biological pathways. Typical nutritional sources wealthy in tocopherol and tocotrienols include vegetable oils, nuts and grain germ part. Broccoli, garlic, onions, vegetable oils, almonds, and walnuts are among the ingredients that have been shown to have positive immune-modulatory impacts. Similarly, garlic, soybean, cabbage, ginger, licorice root extract (*Glycytthizaglabra* root extract) and umbelliferous plants (plants growing or producing under the floor, e.g. carrots) were recognized as foods and herbs with the greatest anti-cancer activity. Citrus includes a host of active phytochemicals as well as a large supply of vitamin C, folic acid, potassium and soluble fibre. Green tea increases the immunity of humor and cells while reducing the risk of certain cancers and cardiovascular diseases. Ginseng, on the other side, enhances macrophage and T-cell manufacturing, natural killer cells and colony forming bone marrow activity. In epidemiological studies, soy bean, garlic, ginger and green tea have been suggested to decrease the incidence of cancer by causing programmed cell death. Soybean extract was shown to avoid polycystic kidney development. Turmeric is most powerful against tumors of the skin (Chacko et al. 2010).

Probiotics and Prebiotics

Probiotics are the beneficial live microorganisms that confer health benefits on the host, e.g., if administered in adequate amounts. *Biofidobacterium, lactobacillus* (LAB), etc.

Prebiotics are the non-digestible food ingredients that stimulate probiotics growth and activity in the digestive system in ways that claim to benefit health. Prebiotics are therefore healthy, non-digestible food components that assist the useful or good bacteria develop and thrive through our digestive system.

It is thought that probiotics safeguard us in two main respects. The first is their role in our digestive tract. Our digestive tracts need a healthy equilibrium between microorganisms that are good and bad. But our lifestyles like poor food decisions, emotional stress, and absence of sleep, over-use antibiotics, other drugs, and environmental impacts can change the equilibrium in favor of bad microorganisms. It filters out and eliminates stuff that can harm it when the digestive tract is healthy, such as damaging microbes, toxins, chemicals, and other waste products. On the flip side, it incorporates the stuff our body requires (food and water nutrients) and absorbs and helps bring them to the cells where they are required. It's not the concept to kill all the microbes. Our bodies need the bad ones as well as the excellent ones. The issue is when you shift your equilibrium to be worse than nice. Diarrhoea, urinary tract infections, muscle pain, and exhaustion were correlated with an imbalance. Probiotics ' second significant advantage is their effect on our immune system. Our immune system is our pathogen protection. We may suffer from allergic reactions, autoimmune illnesses (e.g. ulcerative colitis, Crohn's disease, and rheumatoid arthritis) and diseases (e.g. infectious diarrhoea, Helicobacter pylori, skin infections, and vaginal infections) if it does not work correctly. These ailments are avoided by keeping the right equilibrium from birth. A newborn picks up the beneficial bacteria from his / her mom during delivery through the birth channel. These good bacteria are not transferred when a Caesarean section is done and have been shown to be the reason why allergies, sub-optimal immune systems and reduced concentrations of gut microflora are present in some babies born from Caesarean section. Some of the particular mechanisms by which probiotics exclude undesirable microorganisms include inhibitory substance manufacturing, adhesion site blockage, nutrient competition, toxin receptor degradation, and immunity stimulation. Prebiotics stimulate the development of useful and healthy microorganisms (probiotics) in the intestine, resulting in increased resistance to pathogen invasion. This beneficial effect of prebiotics is known as the prebiotic effect in the human intestine in an unchanged form. However, such prebiotic effects occur when the amount and activity of probiotics increases. A particular microbial group is not emphasized by the prebiotic definition.

The advertisement of dietary supplements and fermented food products containing useful crops. These cultures are what is now to be regarded as probiotics. Cereal juice, frozen yogurt, granola, candy bars, and cookies are other foods presently claimed to provide probiotics. While probiotics may be contained, there is no assurance that they will have them at ideal concentrations. If there are any studies to support his particular allegations, only the product supplier can verify. Non-starch carbohydrates like soluble dietary fibers (e.g. β-glucan, inulin, etc.) and other oligosaccharides such as fructo-oligosaccharides (fructans), galacto-oligosaccharides, etc. are the most common types of prebiotics. Many of the crops commonly consumed as vegetables are great sources of inulin–asparagus, garlic, leek, onion, artichoke. In many crops with nutritional fibers and fructans, β-glucan and inulin are prevalent. Soybeans, raw oats, unrefined wheat and unrefined barley are also traditional nutritional sources of prebiotics. Some of the oligosaccharides that happen naturally in breast milk are thought to play a significant part through the prebiotic-probiotic connection in the growth of a good immune system in children (Dhingra et al. 2012).

Effect of Processing on Functional Components in Food

The processing of a food impacts its functional parts. Some methods of food processing boost the concentration of these functional elements in food while others decrease it. The duration of post-harvest storage, steam blanching, and thermal processing has been shown to all affect the retention of functional compounds in allium vegetables. However, losses of approximately 30–80% of bioactive isothiocyanates have been recorded through thermal processing. Additionally, high temperatures (100 °C and above) inactivate important enzymes, cruciferae myrosinase and allinase in allium vegetables, thereby decreasing the number of functional parts. Temperatures associated with normal cooking, however, have shown little evidence of significant isothiocyanate loss. The liquidation of glucosinolates and their hydrolysis products also leads to a decrease in complete post-cooking phytochemical content. Research has shown that microwave garlic heating for 30–60 s or heating it to 60–100 °C temperature outcomes in important losses of its anti-inflammatory, anti-cancer, antimicrobial, and anti-oxidant operations. The bioavailability of carotenoids and other lipid-soluble functional food elements has been demonstrated to enhance with surface-enhancing processing technique, such as cutting and chopping, as well as heat treatment that breaks down protein and carbohydrate matrix. Tea leaves brewing, whether black or green, releases 69–85% of their bioactive flavonoids in warm water within 3–5 minutes. However, drying pre-treatment before boiling was discovered to decrease cooking time resulting in less leaching of antioxidants in *Homanthalia elongata*, an edible Irish seaweed. Drying as far as extract is concerned, followed by boiling *H. Elongata* had the most important impact on phytochemicals with a 174% increase in its complete phenolic content. However, compared to extracts from new samples, this handling therapy decreased its anti-microbial activity (Xu et al. 2017).

Examples of a Few Common Vegetables with High Quantity of Functional Components

All vegetables contain vitamins, minerals and fiber that are healthy. Some, however, stand out for their outstanding health advantages. Some people, depending on their diet, general health, and nutritional needs, benefit more from certain vegetables than others.

Spinach

Spinach is a vegetable of leafy green. It is also an excellent source of calcium, iron, vitamins, and antioxidants. Spinach is an excellent complement to any meat-or dairy-free diet because of its iron and calcium content. One cup of raw spinach consists mostly of water, containing only seven calories. It also offers nutrients as follows:

- An adult's full daily requirements for Vitamin K
- High amounts of Vitamin A
- Vitamin C
- Magnesium
- Folate
- Iron
- Calcium
- Antioxidants

For a healthy body, and particularly for powerful bones, vitamin K is vital. It increases the way calcium is absorbed by the body. Spinach also offers energy and healthy blood with a good quantity of iron and a good level of magnesium for muscle and nerve function. It's wealthy in antioxidants as well. Research reports that leaves of spinach may reduce blood pressure and benefit the health of the core (Ifolubapo 2018).

Kale

Kale is a famous green leafy vegetable with a number of health advantages. It offers 7 calories per cup of raw leaves and vitamins A, C, and K in decent quantities. Kale can benefit high cholesterol individuals. One tiny research reported experiencing males with elevated cholesterol drinking 150 milliliters of kale juice a day for 12 weeks:

- A 10 percent reduction in low-density lipoprotein, or “bad” cholesterol
- A 27 percent increase in high-density lipoprotein, or “good” cholesterol

Other study indicates blood pressure, blood cholesterol, and blood sugar concentrations are reduced by kale juice.

Broccoli

Broccoli is an unbelievably healthy vegetable belonging to the same family as chicken, kale and cauliflower. All these are vegetables that are cruciferous. Each cup of broccoli sliced and cooked includes:

- 55 calories
- The full day requirement for Vitamin K
- Twice the daily recommend amount of Vitamin C

Eating plenty of cruciferous vegetables can decrease cancer danger, which can be due to sulforaphane contained in these vegetables. Sulforaphane decreased the size and amount of cells of breast cancer and blocked tumor development in animal research.

Peas

Peas are sweet, starchy vegetables which supply 134 calories per cup cooked. They also have a high level of:

- Fiber, at 9 grams per serving
- Protein, at 9 grams per serving
- Vitamins A,C, and K
- Certain B vitamins

Green peas are a useful source of plant-based protein, and eating them is an efficient way to increase your protein consumption for vegetarian and vegan individuals. Peas and other legumes contain fiber that promotes good intestinal bacteria to guarantee periodic intestinal motions and a healthy digestive tract. They are also rich in saponins that can provide antioxidant and anticancer advantages as plant compounds.

Sweet Potatoes

Sweet potatoes are root vegetables which, when cooked with their skin, provide 103 calories and 0.17 grams of fat per medium potato. Each potatoes also contains:

- Much more than an adult's daily requirement of Vitamin A
- 25 percent of their vitamin C and B-6 requirements
- 12 percent of their potassium needs
- Beta-carotene, which may improve eye health and fight cancer

People with diabetes may profit from sweet potatoes. This is because they are small and high in fiber on the glycemic index scale, so they can assist control blood sugar.

Beets

One cup of beets contains 58 calories, along with:

- 442 milligrams of potassium
- 148 micrograms of folate

Beets and beetroot juice are good for heart health improvement. This vegetable is high in nitrates that are healthy to the heart. A 2012 small-scale study claims that drinking 500 grams of beetroot juice in healthy people considerably reduced blood pressure. These vegetables can benefit diabetes individuals as well. Beets contain an antioxidant called alpha-lipoic acid, called diabetic neuropathy, which may be useful for diabetes-related nerve issues.

Carrots

Each cup of sliced carrots includes 52 calories and the daily recommended intake of vitamin A in the form of beta-carotene by an adult more than four times. For good eyesight, vitamin A is essential, and getting enough of this nutrient can help avoid loss of vision. Some carrot nutrients may also have cancer-fighting characteristics. A 2011 research reported that leukemia cells may be killed or inhibited by carrot juice extract.

Fermented Vegetables

Fermented or pickled vegetables provide all of their unfermented counterparts ' nutrients as well as a good dose of probiotics. Probiotics are beneficial bacteria found in the body and in certain foods and supplements. Some scientists think they can enhance the well-being of the gut. Probiotics can help with the symptoms of irritable bowel syndrome, according to the National Center for Complementary and Integrative Health. They can also stop diarrhea caused by infections and antibiotics. Some good vegetables for fermentation include:

- Cabbage, to make sauerkraut
- Cucumbers, to make pickles
- Carrots
- Cauliflower

Tomatoes

Although tomatoes are a fruit technically, they are treated by most individuals as vegetables. Every cup of raw tomato chopped contains:

- 32 calories
- 427 milligrams of potassium
- 24.7 milligrams of Vitamin C

Tomatoes contain lycopene, a strong antioxidant that can play a part in the prevention of cancer. Research indicates lycopene can help avoid cancer of the prostate. Beta-carotene also has anticancer impacts in tomatoes. Vision may be protected by other powerful antioxidants in tomatoes such as lutein and zeaxanthin. The Age-Related Eye Disease Study claims that the risk of age-related macular degeneration has been lowered by individuals who have elevated nutritional intakes of these drugs.

Garlic

In both cooking and medicine, garlic has been used for a long time. Each garlic clove includes only 4 calories and is small in minerals and vitamins. Garlic is an antibiotic that is natural. According to a 2012 laboratory-based research, a compound in garlic, called diallyl sulfide, may be more efficient than two common antibiotics to fight the *Campylobacter bacterium.*

Onions

Each cup of chopped onions provides:

- 64 calories
- Vitamin C
- Vitamin B-6
- Manganese

Onions and other vegetables contain sulfur compounds, including garlic. These can assist to safeguard against cancer. Men with the largest intake of allium vegetables had the lowest prostate cancer risk in other studies. Regular use of onions and other allium vegetables can also decrease the risk of cancer of the esophagus and the stomach.

Alfalfa Sprouts

Each cup of Alfalfa Sprouts contains only 8 calories and a good amount of Vitamin K. These sprouts also boast several plant compounds that contribute to good health, including:

- Saponins
- Flavonoids
- Phytoestrogens

Specialists traditionally use alfalfa sprouts to treat a variety of health circumstances, such as arthritis and kidney issues. However, for these circumstances, scientists have performed little research on their efficacy. Studies in animals have suggested that sprouts of alfalfa may have antioxidant impacts and may decrease inflammation. Eating sprouted legumes can have additional health advantages. Studies indicate that plants that sprout or germinate improve their content of protein and amino acids. Germination can also enhance alfalfa and other seed digestibility and boost dietary fiber content.

Bell Peppers

Sweet bell peppers are frequently found in varieties of red, yellow, or orange. Also common are unripe, green peppers, but they smell less sweet than other colors. A cup of chopped red bell peppers provides:

- 39 calories
- 190 milligrams of Vitamin C
- 0.434 milligrams of Vitamin B-6
- Folate
- Beta-carotene, which the body converts into Vitamin A

Capsanthin, quercetin, and lutein (in the green variety) are antioxidants present in bell peppers that protects vision.

Cauliflower

One cup of chopped cauliflower contains:

- 27 calories
- Plenty of Vitamin C

- Vitamin k
- Fiber

Dietary fiber enhances the heart and guts health of a person, prevents digestive problems and reduces obesity. There is an antioxidant calledindole-3-carbinol (I3C) in cauliflower and other cruciferous vegetables. I3C can decrease breast cancer and breast cancer in both males and females. Like broccoli, another potential anticancer compound is found in cauliflower: sulforaphane.

Seaweed

Seaweed, also known as marine vegetables, is a versatile and nutritious plant that offers many health advantages. Common seaweed kinds include:

- Kelp
- Nori
- Sea lettuce
- Spirulina
- Wakame

Seaweed is one of the few sources of docosahexaenoic acid and eicosapentaenoic acid omega-3 fatty acids based on plants. These fatty acids are vital to the health of a person and are mostly found in sources of meat and dairy. Each sort of seaweed has a slightly distinct dietary profile, but is typically rich in iodine, which is an important thyroid nutrient. Eating a range of sea vegetables offers several significant antioxidants for individuals to decrease harm to the cells of the body. Many kinds of marine algae contain chlorophyll, which is an anti-inflammatory plant pigment. Brown marine vegetables like kelp and wakame contain another powerful antioxidant called fucoxanthin. Research indicates that vitamin E's antioxidant strength is 13.5 times that of fucoxanthin (Sokola et al 2018).

Conclusion

Clearly, due to their medicinal characteristics, functional elements in food will play a significant role in future health maintenance. However, to optimize health advantages, the bioavailability of these functional food parts and the concentrations needed in humans are critical variables. There is inadequate and hazy current data in this respect. Consequently, more data needs to be provided to customers to efficiently guide the making of wider dietary decisions that contain ideal concentrations of functional food parts that promote health.

References

Abuajah CI, Ogbonna AC, Osuji CM. 2015. Functional components and medicinal properties of food: a review. J Food Sci Technol. 52 (5):2522-2529.

Balić A, Vlašić D, Žužul K, Marinović B, Bukvić Mokos Z. 2020. Omega-3 versus Omega-6 Polyunsaturated Fatty Acids in the Prevention and Treatment of Inflammatory Skin Diseases. Int J Mol Sci. 21(3):741.

Chacko SM, Thambi PT, Kuttan R, Nishigaki I. 2010. Beneficial effects of green tea: a literature review. Chin Med. 5: 13.

Cory H, Passarelli S, Szeto J, Tamez M, Mattei J. 2018. The Role of Polyphenols in Human Health and Food Systems: A Mini-Review. Front Nutr. 5:87.

Dhingra D, Michael M, Rajput H, Patil RT. 2012. Dietary fibre in foods: a review. J Food Sci Technol. 49(3):255-266.

El Khoury D, Cuda C, Luhovyy BL, Anderson GH. 2012. Beta glucan: health benefits in obesity and metabolic syndrome. J Nutr Metab. 2012:851362.

Ifeoluwapo Amao. 2018. Health Benefits of Fruits and Vegetables: Review from Sub-Saharan Africa, Vegetables - Importance of Quality Vegetables to Human Health, Md. Asaduzzaman and Toshiki Asao, Intech Open.

Khoo HE, Prasad KN, Kong KW, Jiang Y, Ismail A. 2011. Carotenoids and their isomers: color pigments in fruits and vegetables. Molecules. 16(2):1710-1738.

Rizvi S, Raza ST, Ahmed F, Ahmad A, Abbas S, Mahdi F. 2014. The role of vitamin e in human health and some diseases. Sultan Qaboos Univ Med J. 14(2):e157-e165.

Salehi B, Quispe C, Sharifi-Rad J, et al. 2021. Phytosterols: From Preclinical Evidence to Potential Clinical Applications. Front Pharmacol. 11:599959.

Sokoła-Wysoczańska E, Wysoczański T, Wagner J, et al. 2018. Polyunsaturated Fatty Acids and Their Potential Therapeutic Role in Cardiovascular System Disorders-A Review. Nutrients. 10 (10):1561.

Xu DP, Li Y, Meng X, et al. 2017. Natural Antioxidants in Foods and Medicinal Plants: Extraction, Assessment and Resources. Int J Mol Sci. 18(1):96.

7

Natural Bioactive Compounds in Pulses

[1]Devegowda S R, [2]Saket Kushwaha, [3]V. Kamalavanshi and [4]Shivananda P. Yarazari

[1]Department of Agricultural Economics, Banaras Hindu University Varanasi-221005, Uttar Pradesh
[2]Rajiv Gandhi University, Papum Pare, Itanagar, Arunachal Pradesh-791112
[3]Department of Agricultural Economics, Banaras Hindu University Varanasi-221005, Uttar Pradesh
[4]Department of Extension Education, Banaras Hindu University Varanasi-221005, Uttar Pradesh

Introduction

Pulses are annually grown leguminous crops that yield one to twelve grains/seeds within a pod, and are often promoted in diet owing to their low cost and many beneficial nutritional effects (Craig 2009). The United Nations general assembly has declared 2016 as the international year of pulses (Oyeyinka et al. 2016). The Food and Agriculture Organization (FAO) describes pulses as the crops that are harvested only for dry grain. This definition excludes crops that are harvested green as food sources (such as green beans and green peas), crops that are used only for sowing purposes (such as alfalfa and clover) and crops used primarily for oil extraction (such as peanuts and soybean). India is the largest producer (around 25%) of pulses in the world and the Indian subcontinent covers the leading area in terms of consumption of pulses. It was also suggested that India can also increase the production output by 30%, considering a rise in government support prices for stimulating the growth of pulses and in case of a normal monsoon season (FAOSTAT 2008). The mainly grown pulses in India include mungbean (green gram), chickpea (Bengal gram), urd bean (black gram), lentil (masur), pigeon pea (arhar), cowpea, horse gram and peas (Tiwari and Singh 2012). In the western countries, nowadays pulses are one of the preferred food sources of the population, owing to ideological reasons and interest in vegetarian diet.

Pulses are a good source of nutritional components (Tharanathan and Mahadevamma, 2003), they are low in fat and rich in complex carbohydrates, vegetable protein and minerals phosphate. They are an excellent source of dietary fiber (USDA, 2005; USDHHS, 2015); researchers have found a positive correlation between the anticarcinogenic effects and various components present in pulses, including dietary fiber (Mathers, 2002; Champ, 2002; Howe, 1992) and folate (Rampersaud, 2002). Pulse grains also represent a source of several natural bioactive substances such as lectins, enzyme inhibitors, oxalates, oligosaccharides, phytic acid and phenolic compounds, these substances play important metabolic roles in humans or animals that frequently consume these sensational seeds (Champ, 2002). Scientific studies have shown that consumption of pulses (three or four times per week) can contribute to the prevention of coronary heart disease (CHD) and improve well-being of human (Flight and Clifton, 2006; Bazzano et al., 2001). The mechanisms responsible for this apparently protective role can have a favorable influence on glucose and blood lipids. Also, pulses are a good source of minerals, especially beans and lentils which are very rich on iron (Vadivel and Janardhanan, 2000; MH, 1999; Mataix and Mañas, 1998; Elhardallon and Walker, 1992) and zinc (Vadivel and Janardhanan, 2000; Vadivel et al., 2000; Jimenez et al., 1998; Terrés et al., 2001). The major polyphenolic compounds of pulses are tannins, flavonoids and phenolic acids. The dark grains such as black gram, lentil and red kidney beans are highly rich with the phenolic content. They are organic molecules characterized by the presence of at least two phenolic groups, associated in complex structures, generally of high molecular weight. These compounds are the products of secondary metabolism of plants; they have a good antioxidant activity assessed by 2, 2-diphenyl-1- picryhydrazyl (DPPH) free radical scavenging (Amarowicz and al., 2004; Awika and al., 2003; Xu and Chang, 2007). Some phenolic compounds can decrease protein digestibility (Chung and al., 1998), mineral bioavailability (Sandberg, 2002) and diminish development of various degenerative diseases (Harman, 1956). Phenolic acids are aromatic secondary plant metabolites, they are the main polyphenols made by plants. Flavonoids are a family of polyphenolic; they are phytochemicals compounds found in almost pulses. They have been assayed in various diseases affecting the heart, brain, and other disorders, including those leading to cancer. Flavonoids and phenolic acids have antioxidant and good physiological and biological properties (Murphy and Hendrich, 2002; Pieta, 2000; Yeh and Yen, 2003; Wei and Wang, 2015). Phytic acid (myo-inositol hexaphosphate) is a naturally occurring molecule found in pulses, it can diminish mineral bioavailability (Sandberg, 2002) and exhibit antioxidant activity and protects DNA damage by creating complexes with Fe (Midorikawa et al., 2001; Phillippy, 2003; Graf et al., 1987; Vucenic and Shamsuddin, 2006).

Pulses are also rich in Protein inhibitors of hydrolases active against proteases, amylases, glycosidases, lipases and phosphatases (Belitz and Weder, 1990; Campos- Vega et al. 2010). They are the main sources of lectins in ordinary human food. This small protein that can bind to cell membranes is one of the major proteins found in lentil (Zhang, 2009). Plant lectins present in pulses are important tools in cell biology and immunology (Grant, 1995); a number of studies have suggested an important correlation between certain lectin-binding patterns and their biological behavior in diverse tumors (Jordinson, 1999; Sames, 2001; Wang, 2000; Lorea, 1997; Remmelink, 1999). On the other hand, some enzyme inhibitors and lectins have an antinutritional effects, they can diminish protein digestibility and nutrient absorption, but their effect is effectively remove after cooking (Habiba, 2002; Lajolo and Genovese, 2002). Another important natural compounds found in the seeds and underground organs of pulses (beans, chickpeas, lentils etc), these are short chains of galactose molecules called the Galactooligosaccharides (GOS). Clinical trials have examined the laxative effects of galactooligosaccharides, they have shown that the use of GOS may reduce the severity of constipation in older adults with low defecation frequency (Hamilton-Miller, 2004; Narimiya, 1996; Teuri and Korpela, 1998; Sairanen, 2007; Niittynen, 2007), but in adults with normal bowel function, the effects of GOS was limited (Niittynen, 2007; Deguchi et al., 1997; Teuri et al., 1998) and may cause flatulence (Teuri and Korpela, 1998; Muzquiz et al., 1999). On the other hand, these same molecules can elicit prebiotic activity (De Boever et al., 2000; Rycroft et al., 2001; Macfarlane et al., 2008; Ito, et al., 1993; Gibson et al., 1995) and have protective effects against cancer (Lajolo and Genovese, 2002; Mathers, 2002).

The present research is focusing on replacing energy dense foods with pulses as it has shown to have beneficial effects not only on the management, but also prevention of obesity and other related disorders (Rebello et al., 2014). Polyphenols from pulses are of particular interest as these not only contribute to the seed colour as well as sensory characteristics, but also provide several biological properties including antioxidant, anti-inflammatory, anti-artherogenic and antimicrobial effects (Balasundram et al., 2006). Moreover, phytosterols have emerged as valuable nutraceuticals and functional foods in the market. These compounds are essential components of the lipid membrane bilayer in all plant tissues and show antibacterial, anti-inflammatory, antitumor and anti-ulcerative properties (Uddin et al., 2015). Dietary fibre (8–28% in concentration) as a bioactive component of pulses is another area of interest, which depends on the variety, species and processing methods used. Moreover, other non-nutritional substances in pulses can potentially have beneficial effects in physiological conditions, but the safe as well as the minimum concentrations of these compounds required for the desirable effects need to be determined and afterwards optimized. Therefore, the aim of this review is to provide the current knowledge of certain classes of bioactive compounds present in pulses including polyphenols, phytosterols, resistant starch, dietary fibre and oligosaccharides. Additionally, the antioxidant activity and human health benefits of these metabolites are also discussed.

Polyphenols

Polyphenols have an aromatic ring that bears one or more than one hydroxyl substituents that range from simple to highly polymers compounds (Singh et al., 2015). These compounds are bioactive secondary metabolites widely distributed in all higher plants and are mainly synthesized by the shikimic acid pathway. These phytochemicals are generally categorized into four major groups that include flavonoids, phenolic acids, lignans and stilbenes (Singh et al., 2016c). Bioactive compounds such as polyphenols have a wide therapeutic potential as these can contribute towards the antioxidant activities (Singh et al., 2016a). Phenolic acids and their derivatives flavanols, flavan-3-ols, anthocyanins, condensed tannins are the main polyphenol categories present in legume seeds. Flavonoids, phenolic acids and procyanidins are the dominant phenolic compounds present in lentils, peas and common beans (Zhang et al., 2015). Dehulling of pulses results in removal of large amounts of polyphenols as these are concentrated in the seed coat portions. Girish et al. (2012) reported that about one-fourth of the black gram grain was removed as a by-product after milling process which includes the seed coat rich in polyphenols.

Seed coat of legumes acts as a protective barrier for the cotyledon and it contains high concentration of phenolic compounds (Dueñas et al., 2006). The polyphenols are mainly present in the seed coat portions of beans, while cotyledons contain relatively lower amounts. Flavonol glycosides, anthocyanins and tannins were reported to be responsible for the seed coat colour in dry beans. The amounts of cyanidin 3,5-diglucoside, cyanidin 3-glucoside, delphinidin 3-glucoside, petunidin 3-glucoside, pelargonidin 3-glucoside in seed coats of kidney beans were in the ranges of 0–0.04, 0–0.12, 0–2.61, 0–0.17 and 0–0.59mg/g dry weight [DW] (Choung et al., 2003). Phenolic and tannin contents of four common beans having different seed coat colors were reported to vary significantly (Madhujith et al., 2004). Red, brown and black bean extract contained high polyphenol and tannin content, while white whole bean extract had the lowest amount. Total phenolic content (TPC) varying from 11.2 to 25.3 mg catechin (CE) equivalents/g DW has been reported for cotyledons of seven bean cultivars (Rocha-Guzmán et al., 2007). Luthria and Pastor-Corrales (2006) separated and quantified phenolic acids present in fifteen dry bean varieties using high performance liquid chromatography (HPLC). They detected and quantified ferulic acid (which was also the most abundant), p-sinapic acid and coumaric acid in all of the varieties, while caffeic acid was only detected in two black bean varieties. Lin et al. (2008) reported that hydroxycinnamic acids were similar for commercial bean varieties, but the flavonoid components showed distinct differences. Moreover, black beans were observed to contain primarily 3-O-glucosides of malvidin, petunidin and delphinidin, whereas the main flavonoids in kidney beans were diglycosides of quercetin and kaempferol. Bean coats have a relatively higher level of TPC and flavonoids than most cereal grains and are potential candidates to be used as food preservatives. Jin et al. (2012) characterized proanthocyanidins (oligomeric and polymeric polymeric flavonoids) in lentil, pea and faba bean seeds. In peas, the proanthocyanidin subunit composition was mainly prodelphinidin (gallocatechin and epigallocatechin), while procyanidin- and prodelphinidin-type flavan-3-ol subunits were the main proanthocyanidins of faba beans and lentils.

Aguilera et al. (2011) investigated phenolic profile of Cannellini and Pinta bean flours as well as foods that were bean-based. Several phenolic compounds, non-flavonoids (such as hydroxybenzoic and hydroxycinnamic acids) and flavonoids were identified in the study. The seed coat of lentils was reported to be rich with catechins and procyanidins dimers, while myricetin, quercetin, apigenin and luteolin glycosides are presented in minor concentrations (Dueñas et al.,2006).On the other hand, the cotyledons mainly contain hydroxybenzoic and hydroxycinnamic acids (Dueñas et al., 2006). Han and Baik (2006) reported that TPC of lentils was much higher than that of chickpeas and yellow peas. Amarowicz et al. (2010) identified twenty polyphenolic compounds in the crude extracts of lentils which included flavonols, hydroxycinnamates, procyanidins, dihydrochalcones, dihydroflavonols and gallates using HPLC-PAD and HPLC–MS techniques. Dueñas et al. (2014) elucidated specific polyphenols present in the seed coat and cotyledons of 2 varieties of dark peas using HPLC-PAD-MS. Quercetin, luteolin and apigenin, proanthocyanidins, flavones, flavonol glycosides and trans-resveratrol-3-glucoside were found in the seed coat. The cotyledon contained hydroxybenzoic, hydroxycinnamic, trans pcoumaroyl-malic acid and p-hydroxybenzoyl-malic acid. Mirali et al. (2014) successfully applied HPLC–MS method for analyzing polyphenols from seed coats (green, black and grey) of three lentil genotypes. They reported that amounts of oligomeric flavan-3-ols were almost similar for all the three genotypes, while luteolin as well as its glycosylated form were present in more amounts in the black seed coat genotype. Zhang et al. (2015) identified twenty-one phenolic compounds (with the majority being flavonoids including catechin glucosides, procyanidins and kaempeferol glycosides) from twenty Canadian lentil cultivars (green and red).

Lentil seed coat contains more abundant and diverse type of polyphenols as compared to cotyledon. Different sub-classes (6 in total) of phenolic compounds were found in the seed coat and they mostly consist of oligomers and polymers of proanthocyanidins (Mirali et al., 2014). Phenolic acids, flavan-3-ols, proanthocyanidins, anthocyanidins, stilbenes, flavonols, flavones and flavanones are the major polyphenols identified in lentil seeds (Alshikh et al., 2015). TPC in free, esterified and insoluble forms were reported in the range of 1.37–5.53, 2.32–21.54 and 2.55–17.51mg GAE/g of defatted lentil samples, respectively (Alshikh et al., 2015). Free and esterified flavonoids in lentils were observed in the range of 0.01–0.80 and 0.36–4.13 mg catechin eq/g defatted sample, respectively. Aguilera et al. (2011) identified 35 phenolic compounds in raw and processed lentil flours. The identified phenolic compounds were catechins and procyanidins (69%), flavonols (17%), flavones and flavanones (5%), hydroxybenzoic (5%) and hydroxycinnamic compounds (4%). High concentrations of flavonoids were reported in the seed coat of lentils. López-Amorós et al. (2006) determined phenolic compounds in raw pea and reported the concentration (μg/100.g DW) of protocatechuic acid (206–221), p-hydroxybenzoic acid (11.9–13.7), vanillic acid (31.2–34.4), trans p -coumaric acid (37.7–41.5), cis p-coumaric acid (65.5–70.1), cis p-coumaric acid derivative (31.9–33.7) and trans ferulic acid (9.9–10.9). Parmar et al. (2016) reported that ferulic acid was the main compound in raw grains of various pulses. Moreover, they reported that raw kidney beans had a greater accumulation of chlorogenic acid, catechin, ferulic acid and p-coumaric acid than field pea and chickpea.

TPC in hull, whole, and dehulled dry beans was observed in the range of 2.2–78.2mg CE/g with most of phenolics are concentrated in the hull portion (56mg CE/g) of beans (Cardador-Martinez et al., 2002). TPC in the crude extracts of adzuki bean, green lentil, red lentil, red bean and pea was 90, 68, 58, 55 and 23mg GAE (gallic acid equivalents)/g DW, respectively (Amarowicz et al., 2010). Total flavonoid content in seeds of pea, chickpea, lentil, red kidney and black bean was in the range of 0.08–3.21mg CAE/g DW (Xu et al., 2007). Lentils (47.6mg/g DW) and kidney beans (45.7mg/g DW) had the highest phenolic content and baby lima bean had the lowest (9.5mg/g DW). The content of catechin, procyanidin B2, B3 and procyanidin tetramer in raw lentils was in the range of 0.1–0.5mg/100.g DW (López-Amorós et al.,2006). Sharma et al. (2015) reported an average phenolic content of 78.9mg GAE/g for field pea. Parmar et al. (2014) documented that dark coloured kidney beans had more TPC than light coloured beans (ranging from 0.1 to 7.1mg GAE/g

DW). Segev et al. (2010) reported that coloured chickpea seeds contained up to 13 and 31 fold more TPC and antioxidant activity, respectively, than beige and regular cream coloured seeds, indicating their use as potential functional foods. Campos-Vega et al. (2010) reported that chickpeas had varying amounts of genistein, formononetin and daidzein (0.06, 0.14 and 0.04mg/100.g DW), respectively.

Phytosterols

Phytosterols are the plant sterols having a structure similar to cholesterol. These compounds are extensively studied as functional food ingredients and well known for their wide variety of health benefits such as lowering of blood cholesterol levels and reducing the intestinal absorption of dietary and endogenous cholesterol. Phytosterols are not synthesized in our body and are taken from the diet by intestinal absorption. These bind with bile acids and prevent their reabsorption. Studies have indicated that pulses have good amount of phytosterols (as a component of plant cell membranes). These bioactive compounds are found in seeds of many legumes like chickpeas, beans, peas, lentils and lupins (Table 1). Phytosterol concentration in kidney beans and chickpeas was reported to be 127 and 35mg/100 g, respectively (Weihrauch and Gardner.1978). Ryan et al. (2007) studied the total phytosterol content in butter beans, chickpeas, kidney beans, lentils and peas and reported it in the range of 134mg/100.g DW (kidney beans) to 242mg/100g DW (peas). In addition, chickpeas and lentils had a phytosterol content of 205 and 158mg/100g DW, respectively. They further measured the β-sitosterol, campesterol, stigmasterol content and reported that the β-sitosterol was the dominant phytosterol in all the studied legumes. Pulses mainly contain phytosterols in range of 124–135mg/100g DW (Weihrauch and Gardner 1978). The phytosterol profile of cooked dry beans, chickpeas and peas were studied by Kalogeropoulos et al. (2010). β-sitosterol, campesterol, stigmasterol and Δ^5-avenasterol were found in cooked dry legumes. β-sitosterol was predominant in all pulses with its highest concentration in chickpeas (38.52mg/100g DW). Their study showed that cooking caused a reduction in phytosterol levels of legumes.

Pulses have a high percentage of glycosylated sterols. Nyström et al. (2012) studied the contents of total sterols and glycosylated sterols in adzukibeans, black beans, black eye beans, red beans, white beans, chickpeas and lentils. Sitosterol, stigmasterol, and campesterol are the common sterols present in these pulses. Lentil contains a high proportion of sitosterol and low proportion of stigmasterol, whereas it is opposite in case of beans. Nyström et al. (2012) reported that most of the beans and lentils had glycosylated sterols ranging between 100 and 150 mg/g DW. Phytosterols in pulses occur in the form of free sterols, acylated steryl glycosides and steryl glycosides. Phytosterols from black bean seed coats were identified and quantified in free and conjugated forms by Chávez-Santoscoy et al. (2014). They identified stigmasterol (as free); campesterol (as free), acylated steryl glycoside and steryl glycoside forms; sitosterol (both free and as acylated steryl glycoside); and Δ^5-avenasterol (as free and steryl glycoside forms). They further reported that the most abundant free phytosterols found in the black bean seed coat extract was sitosterol. Iriti et al. (2009) documented that sitosterol was the major phytosterols component found in beans (27.2 mg/100g DW). These studies indicated that lentils, chickpeas, beans and peas are the main source of phytosterols among pulses.

Bioactive Carbohydrates

Carbohydrates form major portion of pulses accounting to as high as 70% of seeds. Based on the nutritional and physiological roles, these are classified into available (hydrolyzed by gastrointestinal enzymes, absorbed and metabolized by the body) and unavailable carbohydrates (not hydrolyzed by the enzymes and pass into the large intestine). The lower glycemic index (GI) of pulses (slow break down of carbohydrate and slow release of glucose into the bloodstream) is attributed to higher contents of indigestible/unavailable carbohydrates such as dietary fibres, resistant/indigestible starch and some oligosaccharides. These collectively stimulate the activity of prebiotics *Bbifidobacteria*.and. *Lactobacilli*) in the colon, resulting in increased production of short chain fatty acids (SCFA; mainly acetic, propionic and butyric acids) along with organic acids (mainly lactic, succinic and pyruvic) and gases (CH_4, CO_2 and H_2), conferring various health benefits to humans. The SCFA are the preferred respiratory fuel for the cells of colon lining and further contribute to increase colonic blood flow, maintenance of low pH, and prevention of abnormal colonic cell growth (Nugent 2005). The indigestible fractions from black bean and lentil were observed to be the best substrate for the production of SCFA (particularly butyric acid) by Hernández-Salazar et al. (2010).

Dietary Fibre

Dietary fibres are regarded as one of the most important bioactive constituents of pulses. There are several definitions of dietary fibre some include starch polysaccharides while others do not. Tharanathan and Mahadevamma (2003) described dietary fibres as the macromolecules that cannot be digested by human endogenous enzymes and are essentially composed of plant cell wall components. According to the American Association of Cereal Chemists International (AACCI) dietary fibre is the edible part of plants and analogous carbohydrates that resist the digestion and absorption processes in the human small intestine with partial or complete fermentation in the large one (Rebello et al., 2014). Therefore, dietary fibres do not constitute a defined chemical group but represent a combination of heterogeneous substances viz. cellulose, hemicelluloses, gums, pectins, mucilages, β-glucans, lignins, resistant starch, non-digestible oligosaccharides, and other associated plant substances. The dietary fibres are classified into soluble and insoluble dietary fibres (SDF and IDF, respectively) based on their solubility in water and buffer systems. Both seed coat and cotyledon are fibre rich parts of pulse seeds (Parihar et al., 2016). The seed coat is rich in water-insoluble polysaccharides, primarily cellulose, whereas the cotyledon fibre comprises of hemicelluloses, pectins and cellulose having varying degrees of solubility. Pulses are good source of dietary fibre and contain both SDF and IDF, the later being the major portion of pulse dietary fibres (Wang and Toews 2011). Pulse SDF constitutes gums, pectins, fructans, inulins, and some hemicelluloses, whereas IDF included cellulose, some hemicelluloses, lignins, and arabinoxylan (Rebello et al. 2014). The total dietary fibre (TDF) content of different pulses (beans, chickpeas, lentils, and peas) ranged from 14 to 32%, with SDF and IDF ranging from 0–9 and 10–28%, respectively against 3.00–15.02, 0.86–4.33, and 2.14–10.79%, respectively for cereals like wheat, rice and barley (Yadav et al., 2010a; Rebello et al., 2014; Dueñas et al., 2016).

The interaction of fibres with other bioactive constituents e.g. polyphenol also influences the physiological properties and health benefits of dietary fibres. The pulse fibres can interact with the polyphenolic compounds present in pulses through the formation of hydrogen and hydrophobic linkages between polyphenols and cell wall components. The fibres from beans and lentils were reported to contain associated hydroxybenzoic and hydroxycinnamic compounds, flavan-3-ols, procyanidins, flavonols and flavones, wherein IDF contain more associated phenolic compounds than SDF (Dueñas et al., 2016).

Resistant Starch

Starch, a storage carbohydrate, constitutes a major proportion (45–65% of dry matter and 70–75% of total carbohydrates) of almost all pulses. Although it is the primary available carbohydrate in pulses, not all of it is digested and metabolized in the body. Based on the rate of enzymatic digestion, glucose release, and absorption in the gastrointestinal tract, starches are classified into rapidly digestible starch (RDS), slowly digestible starch (SDS), and resistant starch (RS) (Englyst et al., 1992). RDS is rapidly and completely hydrolyzed in the small intestine and is associated with high risk of type-2 diabetes and obesity. SDS is slowly and completely digested and possesses a controlling effect on the blood glucose level. In contrast, RS is not digested in the small intestine by endogenous enzymes consequently pass to large intestine where it functions physiologically similar to dietary fibres as it increases faecal bulk, reduces colonic pH, lowers serum cholesterol and triglycerides levels (Fuentes-Zaragoza et al., 2010) and reduces postprandial glycemic responses. Similar to dietary fibres, RS also functions as prebiotics and promotes the growth of probiotics in human gut. However, the fermentation of RS produces higher amount of butyrate as compared to other indigestible carbohydrates.

The RS content of different pulses (beans, chickpeas, lentils, and peas) ranged from 3.95 to 5.09% against 1.42–2.85%, respectively for cereals like wheat, rice and barley (Yadav et al., 2010a). The higher RS for pulses is attributed to higher amylose content (30–60%), absence of pores on the granule surface, C-type crystallinity, strong interactions between amylose chains and lesser tendency to completely lose crystalline- and granular-structure during cooking (Hoover et al., 2010; Singh 2011). Although amylose is a linear polymer of glucose, it is generally digested at slower rate than highly branched amylopectin owing to lesser relative surface area per molecule (Shevkani et al., 2016). In addition, high amylose content of pulse starches results in increased retrogradation, making these more resistant towards amylolysis by the digestive enzymes (Singh 2011). Additionally, the non-starch carbohydrates present in pulse seeds further account for the lower digestibility of pulse starches. As SDF could increase viscosity of the system, these decrease transit time, reducing the time available for starch digestion in the small intestine. The digestibility of whole pulses (beans, lentils, peas and chickpeas) starch after cooking depended on the integrity of the cell walls which prevented degradation of the starches by digestive enzymes as these remained partially crystalline during cooking where the cell walls were not disrupted and the pulses (navy and pinto beans) with more viscous SDF had higher degree of crystallinity in the cell walls than other pulses. Moreover, the protein-starch interactions could further contribute to lesser digestibility of pulse starches (Rebello et al., 2014). Cooking generally reduces RS content; however the extent of this reduction varies with cooking methods. Yadav et al. (2010b) reported lower content of RS in pressure cooked pulses as compared to boiled counterparts attributing to more uniform and complete gelatinization.

Oligosaccharides

Pulses are known to contain the highest contents of some of the oligosaccharides among all the crops. Total oligosaccharides in different pulses, such as lentils, chickpeas and peas were observed to range between 70.7mg/g DW in yellow peas and 144.9mg/g DW in chickpeas (Han and Baik 2006). Sucrose, stachyose and verbascose were predominant oligosaccharides in beans. However, major pulse oligosaccharides are α-galactosides which are characterized by the presence of α (1 → 6) links between the galactose moieties. These are the second most abundant soluble carbohydrates in the plant kingdom after sucrose. Raffinose (trisaccharide), stachyose (tetrasaccharide) and verbascose (pentasaccharide) are predominant α-galactosides present in pulses. Moussou et al. (2016) reported that lentils have the lowest content of raffinose family with 37.5 mg/g DW compared to faba beans (52mg/g DW), beans (60.9mg/g DW) and peas (66.3mg/g DW). They also reported that cotyledons had higher content of these oligosaccharides than seed coats. Other group of α-galactosides in pulses is galactosyl cyclitols, the most common being a trisaccharide ciceritol (Han and Baik 2006). The α-galactosides are not absorbed or hydrolysed by the human digestive system hence fermented by colonic bacteria, resulting in production of flatulent gases responsible for bloating and abdominal discomfort. However, these help in normalizing bowel function, increasing *Lactobacilli* and *Bifidobacteria* and decreasing. *Enterobacteria* in the colon, and reducing the levels of potentially carcinogenic N-nitroso compounds in the gut (Van Loo 1998). Raffinose and stachyose were reported to have health beneficial physiological effects similar to those of dietary fibres (Berrios et al. 2010).

Anti-oxidant Activity

Numerous studies have suggested that oxidative stress is involved in health problems such as diabetes, obesity, neural disorders, cancer and cardiovascular diseases. Polyphenols are the primary antioxidants that are based upon the donation of hydrogen atoms to the free radicals (Singh et al. 2016a). They delay or prevent oxidation of lipids, proteins and DNA by reactive oxygen species that is produced in cells during oxidation (Amarowicz et al. 2010). Information regarding antioxidant activity of bioactive compounds present in pulses helps in understanding the role of pulses in human diet to reduce the incidence of chronic diseases. Pulses exerted high antioxidant capacity as determined by DPPH, ORAC and FRAP assays (Xu and Chang 2009). López-Amorós et al. (2006) reported that peas and beans undergo a significant increase in antioxidant activity after germination due to changes in their phenolic composition. Similarly, Moussou et al. (2016) reported an increase in antioxidant activities of different pulse flours with toasting. Amarowicz and Pegg (2008) reviewed antioxidant activity and radical scavenging capacity of pulses. They correlated polyphenols of pulses with antioxidant activity and reduced incidence of chronic diseases. Lentils, chickpeas and kidney bean have shown high antioxidant potential in many studies. Amarowicz et al. (2010) investigated the antioxidant and antiradical activities of green lentil seed extract and its isolated low-molecular-weight phenolics and tannin fractions. Phenolic compounds and its fractions have demonstrated antioxidant and antiradical activities.

Antioxidant activity of pulses is related to concentration of polyphenols. Wang et al. (2016) evaluated antioxidant activities of some selected beans from China. They reported positive correlations between total antioxidant activities and TPC. Moreover, they reported that black bean, spring bay bean and pearl bean showed the strongest antioxidant activities among the analyzed bean varieties. Raw lentils have shown high values of the antioxidant activity (66.9.μmol trolox/g DW) as compared to common chickpeas and peas due to high content of phenolic compounds (Aguilera et al., 2011). Gallic acid is a potent antioxidant present in lentils and it accounts for 95% of total hydroxybenzoic acids in black lentils. Flavonoids are the main components among phenolics that contributed towards antioxidant potential of pulses. Flavonoids (cyanidin 3-O-β-D-glucoside, pelargonidin 3-O-β-D-glucoside, delphinidin 3-O-β-D-glucoside) present in beans with red and black seed coat have shown strong antioxidative activity (Tsuda et al.,1994). Seed coats of pulses are rich source of flavonoids. Differences have been observed in the antioxidant capacity of seed coat and cotyledon of lentils and dark peas. The seed coats have presented higher antioxidant activity than the cotyledons in lentils and peas and it is related to high flavonoid content (Dueñas et al., 2006).

Pulses with coloured coats have more content of flavonoids and possess strong antioxidant activities. Natural polyphenols in foods are free radical terminators and quench singlet oxygen (Singh et al., 2016b). Cardador-Martínez et al. (2002) demonstrated that red, brown and black beans possess high antioxidative activity as compared to white beans. Many studies have reported that pulses with a dark seed coat show high antioxidant potential owing to large amounts of phenolic compounds such as proanthocyanidins. Madhujith et al. (2004) demonstrated that coloured beans (red, brown and black) possess high antioxidative activity as compared to white beans due to the presence of condensed tannins. The antioxidant activity of lentil cultivars having different seed coat colours was determined with respect to their total phenolic, flavonoid, and condensed tannin (proanthocyanidin) contents (Alshikh et al., 2015). The flavonoid compounds such as anthocyanins, quercetin glycosides and condensed tannins present in seed coat of beans have higher antioxidant activity than simple phenolics (Beninger and Hosfield 2003). Antioxidant potential of bioactive compounds present in legumes have role in controlling chronic diseases like cancer, ageing and diabetes.

Among different pulses studied by Zhao et al. (2014), lentils have shown the high antioxidant capacity (721 U [unit of total antioxidant activity]/g), DPPH assay (38.5%) and reducing power, while baby lima bean and navy bean showed the lowest values. Chickpea (648 U/g), small red kidney bean (622 U/g) and black kidney bean (602 U/g) are also good in antioxidant activity. Chaieb et al. (2011) determined antioxidant activities of thirteen faba bean genotypes grown in Tunisia. They reported that faba beans were a good source of natural antioxidants independently to their genotype and a significant correlation among the phenolic content and the DPPH scavenging capacity was found. Antioxidant activities of commonly consumed pulses have been reported by many research groups (López-Amorós et al., 2006; Xu and Chang 2009; Amarowicz and Pegg.2008; Han and Baik.2006; Chaieb et al., 2011; Fratianni et al., 2014; Alshikh et al., 2015; Zhang et al., 2015). The results of studies have indicated that pulses can be used as source of natural antioxidants. Some studies have shown that processing and cooking may reduce the antioxidant activity of pulses. Xu and Chang (2009) determined the effect of boiling as well as steaming under high pressure on phenolic content and antioxidant activities of black beans and pinto beans. They reported that in comparison with the original raw beans, both boiling and steaming caused a significant reduction in the TPC, condensed tannin content, total flavonoid content and antioxidant activities (as determined by DPPH, ORAC and FRAP assays).

Pulses contribute towards dietary source of natural antioxidants as they contains significant amount of phenolics and flavonoids (Xu and Chang. 2009). Diet rich in pulses may be useful in prevention or treatment of health problems linked with free radical production. Studies have correlated health benefits of consuming pulses with phenolic content of the pulses. Polyphenols are known for their protective effects on oxidative stress-induced diseases and various physiological activities of the body (Aguilera et al., 2011). These compounds have role in promotion of health and prevention of many diseases. Phenolic compounds play role in specific biological functions by modulating the activity of enzymes and cell receptors. In human diet, polyphenols have role in prevention of cancer, cardiovascular and neurodegenerative diseases associated with oxidative stress (Manach et al., 2004). Siah et al. (2012) reported that crude phenolic extracts from raw as well as roasted faba beans had potential health benefit properties such as potent antioxidant activities (based on in vitro and in vivo assays) and chemo-preventative effects. The occurrence of bioactive phenolic compounds and their antioxidant activity/radical-scavenging capacity make pulses a very useful food for daily inclusion in the human diet (Zhao et al., 2014). Zhang et al. (2015) reported that polyphenols present in lentils not only contributed to the antioxidant activities, but they also inhibited glucosidase and lipase (enzymes which are associated with glucose and lipid digestion), contributing to controlling obesity and blood glucose levels in humans. Moreover, polyphenols also function as anti-allergic, antimicrobial, anticancer and anti-inflammatory agents (Daglia 2012). Pea, faba bean and lentil are good dietary source of polymeric flavonoids (Jin et al., 2012). The polymeric flavonoids have health beneficial effects like antioxidant, anti-diabetic, anti-carcinogenic, and anti-inflammatory activities. They also contribute in promoting vascular health and improving serum lipoproteins.

Phytosterols are also known for many health promoting activities. Phytosterols reduces cholesterol levels by inhibiting its absorption in the intestinal lumen through the inhibition of cholesterol micellar solubility (Chávez-Santoscoy et al.,2014). Phytosterols intake of around 3.g per day significantly decreases the total cholesterol level by 10–15% and is very effective in reducing serum LDL cholesterol levels (Nyström et al., 2012). Dietary sources of phytosterols usually come from pulses, corn and some plant oils. Plant sterols may attennuate the inflammatory activity of immune cells and prevent immunological diseases (Bouic 2001). They are known for anti-cancer, anti-inflammatory, anti-oxidative and immune modulatory activities (Bartnikowska 2009). Phytosterols affect many metabolic processes. They play an important role in the hypocholesterolaemic effects of pulses. Hyperlipidemia is involved in the development of several chronic diseases. Phytosterols have important role in reducing the risk of cardiovascular diseases by lowering absorption of lipids at intestinal level in people consuming pulses as food (Kushi et al.,1999; Chávez-Santoscoy et al., 2014). Phytosterols, dietary fibre and resistant starch present in pulses have hypolipidemic benefits (Ramírez-Jiménez et al., 2015).

The SCFA produced by bean fibre fermentation were associated with the regulation of appetite and satiety, acetic and propionic acids influenced serum lipid profiles (Ramírez-Jiménez et al., 2015). SCFA, in particular butyrates, exhibited potent anti-tumour and anti-inflammatory activities while propionate decreased serum cholesterol levels, inhibited fatty acid synthesis and induced satiety (Lanza et al.,2006). The indigestible fractions from black bean and lentil were observed to be the best substrate for the production of SCFA (particularly butyric acid) by Hernández-Salazar et al. (2010). However, McBurney and Thompson (1987) showed that the fermentation of whole black bean resulted in the production of higher levels of acetic acid than propionic and butyric acids.

Dietary fibres present in pulses have beneficial effects in many human health problems. The main properties relating to health benefits of SDF are viscosity, water holding capacity (WHC), organic compound entrapment, and fermentability. SDF are more easily fermented in the large intestine than IDF. Some SDF increase viscosity of bolus resulting in stomach distension, delayed gastric emptying rate, slower intestinal transition and reduced absorption rate of nutrients, resulting in increased satiety and controlled postprandial glucose and lipids levels (Rebello et al., 2014), making them a good choice for people with type-2 diabetes, obesity and cardiovascular diseases. Furthermore, the SDF had the ability to bind bile acids, leading to reduced absorption of dietary cholesterol (Kahlon et al., 2005). Therefore, the consumption of 5–10.g soluble fibre/day was reported to reduce low-density lipoprotein (LDL) cholesterol by approximately 5% (Panel 2002). IDF are fermented slowly and do not form viscous gels (contrary to that of SDF). However, IDF act as bulking agent in the intestine, increasing both the size and weight of food remnants (due to their high water binding capacity and resistance to degradation by bacteria), resulting in laxation. IDF act primarily in the large intestine where these increases faecal bulk, dilutes colonic contents and decreases mouth-to-anus transit time. Therefore, pulses with high content of IDF could also be helpful in regulating bowel movements which were associated with the reduced risk of colon cancers (Costa et al., 2006). This review suggested that pulses are carriers of several bioactive constituents of potential biological importance which contributes to human health and wellbeing.

Conclusion

People nowadays face many health problems, which can be prevented or treated by increasing consumption of nutritional foods that rich in beneficial bioactive constituents. Pulses are important part of human diet in many countries of the world. Pulses are low-fat, dry seeds of legumes such as beans, peas, chickpeas, and lentils. Pulses not only provide significant amount of proteins, carbohydrates and micronutrients but are also rich in bioactive constituents like polyphenols, phytosterols, resistant starch, oligosaccharides and dietary fibre that are beneficial for managing and preventing chronic diseases. Pulses have a low glycemic index and are rich source of dietary fibre and antioxidants. Including pulses in our diet can contribute to prevent or cure diabetes, heart problems, obesity and certain types of cancers. Important biological activities have now been sug- gested for these bioactive compounds like enhancement of the antioxidant, antimutagenic, anticarcinogenic and anti-hyperglycemic effects, which makes pulses an important crop for human health.

References

Adebamowo CA, Cho E, Sampson L, Katan MB, Spiegelman D, Willett WC, Holmes MD. Dietary flavonols and flavonol-rich foods intake and the risk of breast cancer.Int J Cancer.2005; 114:628–633.

Aguilera Y, Estrella I, Benitez V, Esteban RM, Martín-Cabrejas MA. Bioactive phenolic compounds and functional properties of dehydrated bean flours.Food Res Int.2011; 44:774–780.

Alshikh N, de Camargo AC, Shahidi F. Phenolics of selected lentil cultivars: antioxidant activities and inhibition of low-density lipoprotein and DNA damage.J Funct Foods.2015; 18:1022–1038.

Amarowicz R, Estrella I, Hernández T, Robredo S, Troszyńska A, Kosińska A, Pegg RB. Free radical-scavenging capacity, antioxidant activity, and phenolic composition of green lentil (*Pisum sativum*).Food Chem.2010; 121:705–711.

Amarowicz R, Pegg RB. Legumes as a source of natural antioxidants.Eur J Lipid sci Tech.2008; 110:865–878.

Anderson JW, Major AW. Pulses and lipaemia, short- and long-term effect: potential in the prevention of cardiovascular disease. Brit J Nutr. 2002; 88: S263–S271.

Balasundram N, Sundram K, Samman S. Phenolic compounds in plants and agri-industrial by-products: antioxidant activity, occurrence, and potential uses. Food Chem. 2006; 99:191–203.

Bartnikowska E. Biological activities of phytosterols with particular attention to their effects on lipid metabolism.Pol J Food Nutr Sci.2009; 59:105–112.

Bazzano LHJ, Ogden LG, Loria C, Vupputuri S, Myers L, Whelton P K. Legume consumption and risk of coronary heart disease in US men and women: NHANES I Epidemiologic Follow-up Study. Archives of Internal Medicine. 2001; 161:2573-2578.

Belitz HD, Weder JKP. Protein inhibitors of hydrolases in plants foodstuffs. Food Reviews International 1990:6:151-211.

Beninger CW, Hosfield GL. Antioxidant activity extracts, condensed tannin fractions, and pure flavonoids from.*Pisum sativum*.L. seed coat color genotypes.J Agric Food Chem.2003; 51:7879–7883.

Berrios JDJ, Morales P, Cámara M, Sánchez-Mata MC. Carbohydrate composition of raw and extruded pulse flours.Food Res Int.2010; 43:531–536.

Bouic PJ. The role of phytosterols and phytosterolins in immune modulation: a review of the past 10 years.Curr Opin Clin Nutr Metab Care.2001; 4:471–475.

Campos-Vega R, Loarca-Piña G, Oomah BD. Minor components of pulses and their potential impact on human health.Food Res Int.2010; 43:461–482.

Campos-Vega R, Oomah BD, Loarca-Piña G, Vergara-Castañeda HA. Common Beans and Their Non-Digestible Fraction: Cancer Inhibitory Activity-An Overview.Foods.2013; 2:374–392.

Campos-Vega R, Loarca-Pina G, Oomah BD. Minor components of pulses and their potential impact on human health. Food Research International.2010; 43:461-482.

Cardador-Martinez A, Loarca-Pina G, Oomah BD. Antioxidant activity in common beans (*Pisum sativum*.L.).J Agric Food Chem.2002; 50:6975–6980.

Chaieb N, González JL, López-Mesas M, Bouslama M, Valiente M. Polyphenols content and antioxidant capacity of thirteen faba bean (*Pisum sativum*.L.) genotypes cultivated in Tunisia.Food Res Int.2011; 44:970–977.

Champ MMJ. Non-nutrient bioactive substances of pulses.Brit J Nutr.2002; 88:307–319.

Chávez-Santoscoy RA, Tovar AR, Serna-Saldivar SO, Torres N, Gutiérrez- Uribe JA. Conjugated and free sterols from black bean (*Pisum sativum*.L.) seed coats as cholesterol micelle disruptors and their effect on lipid metabolism and cholesterol transport in rat primary hepatocytes.Genes Nutr.2014; 9:367.

Choung MG, Choi BR, An YN, Chu YH, Cho YS. Anthocyanidin profile of Korean cultivated kidney bean (*Pisum sativum*.L.).J Agric Food Chem.2003; 51:7040–7043.

Chung KT, Wong TY, Wei CI, Huang YW, Lin Y. Tannins and human health: A review. Critical Reviews in Food Science and Nutrition.1998; 38:421-464.

Clemente, A., Domoney, C., 2001. Anticarcinogenic activity of protease inhibitors in legumes. Proceedings of the 4th European Conference on Grain Legumes, Cracow: AEP Editions. Paris, pp. 114-115.

Costa GE, da Silva Queiroz-Monici K, Reis SMPM, de Oliveira AC. Chemical composition, dietary fibre and resistant starch contents of raw and cooked pea, common bean, chickpea and lentil legumes.Food Chem.2006; 94:327–330.

Craig WJ. Health effects of vegan diets.Am J Clin Nutr.2009; 89:1627S–1633S.

Daglia M. Polyphenols as antimicrobial agents.CurrOpin Biotech.2012; 23:174–181.

De Boever P, Deplancke B &Verstraete W. Fermentation by gut microbiota cultured in a simulator of the human intestinal microbial ecosystem is improved by supplementing a soygerm powder. Journal of Nutrition.2000; 130: 2599-2606.

Deguchi Y, Matsumoto K, Ito A, Watanuki M. Effects of b1-4 galacto-oligosaccharides administration on defecation of healthy volunteers with constipation tendency. Japanese Journal of Nutrition and Dietetics. 1997; 55:13-22.

Dilawari J, Kumar V, Bhatnagar R, Dash R. Effect of legumes on blood sugar in diabetes mellitus.Indian J Med Res.1987; 85:184–187.

Dueñas M, Estrela I, Hernández T. Occurrence of phenolic compounds in the seed coat and the cotyledon of peas (*Pisum sativum*.L.).Eur Food Res Technol.2014; 219:116–123.

Dueñas M, Hernández T, Estrela I. Assessment of in vitro antioxidant capacity of the seed coat and the cotyledon of legumes in relation their phenolic contents.Food Chem.2006; 98:95–103.

Dueñas M, Sarmento T, Aguilera Y, Benitez V, Mollá E, Esteban RM, Martín-Cabrejas MA. Impact of cooking and germination on phenolic composition and dietary fibre fractions in dark beans (*Pisum sativum*.L.) and lentils (*Pisum sativum*.L.).LWT Food Sci Technol.2016; 66:72–78.

Elhardallon SB, Walker AF. Binding of iron by three starchy legumes in the presence of iron alone, with calcium or with calcium, zinc, magnesium and copper. Journal of Food Sciences and Nutrition.1992; 43:61-68.

Englyst HN, Kingman SM, Cummings JH. Classification and measurement of nutritionally important starch fractions. Eur J Clin Nutr. 1992; 46:S33–S50.

FAO, Food and Agriculture Organization of the United Nations, 1994. Pulses and derived products, in Definition and classification of commodities (draft), Rome.

FAOSTAT (2008) Food & Agric. Organization of the United Nations, Rome, Italy, Year Book January 2008

Flight I, Clifton P. Cereal grains and legumes in the prevention of coronary heart disease and stroke: A review of the literature. European Journal of Clinical Nutrition.2006; 60: 1145-1159.

Fratianni F, Cardinale F, Cozzolino A, Granese T, Albanese D, Di Matteo M, Nazzaro F. Polyphenol composition and antioxidant activity of different grass pea (Lathyrus sativus), lentils (*Pisum sativum*), and chickpea (Cicer arietinum) ecotypes of the Campania region (Southern Italy).J Funct Foods.2014; 7:551–557.

Fuentes-Zaragoza E, Riquelme-Navarrete MJ, Sánchez-Zapata E, Pérez-Álvarez JA. Resistant starch as functional ingredient: a review.Food Res Int.2010; 43:931–942.

Gdala JL, Buraczewska L, Grala W. The chemical composition of different types and varieties of pea and the digestion of their protein in pigs. Journal of Animal Feed Science.1992; 1:71-79.

Gibson GR, Beatty ER, Wang X, Cummings JH.Selective stimulation of bifidobacteria in the human colony by oligofructose and inulin. Gastroenterology.1995; 108:975-982.

Girish TK, Pratape VM, Rao UP. Nutrient distribution, phenolic acid composition, antioxidant and alpha-glucosidase inhibitory potentials of black gram (Vigna mungo.L.) and its milled by-products.Food Res Int.2012; 46:370–377.

Gonzalez De Mejia E, Prisecaru VI. Lectins as bioactive plant proteins: A potential in cancer treatment. Critical Reviews in Food Science and Nutrition.2005; 45:425-445.

Graf E, Empson KL, Eaton JW. Phytic acid. A natural antioxidant. Journal of Biological Chemistry.1987; 262:1647- 1650

Hamilton-Miller JM. Probiotics and prebiotics in the elderly. Postgraduate Medical Journal.204; 80:447-51.

Han IH, Baik BK. Oligosaccharide content and composition of legumes and their reduction by soaking, cooking, ultrasound, and high hydrostatic pressure.Cereal Chem.2006; 83:428–433.

Harman D. Aging: a theory based on free radical and radiation chemistry. Journals of Gerontology.1956; 11:298- 300.

Hernández-Salazar M, Osorio-Diaz P, Loarca-Piña G, Reynoso-Camacho R, Tovar J, Bello-Pérez LA. In vitro fermentability and antioxidant capacity of the indigestible fraction of cooked black beans (*Pisum sativum*.L.), lentils (*Pisum sativum*.L.) and chickpeas (*Cicer arietinum*. L.). J Sci Food Agric.2010; 90:1417–1422.

Hoover R, Hughes T, Chung HJ, Liu Q. Composition, molecular structure, properties, and modification of pulse starches: a review.Food Res Int.2010; 43:399–413.

Howe GR et al. Dietary intake of fiber and decreased risk of cancers of the colon and rectum: evidence from the combined analysis of 13 case-control studies. Journal of the National Cancer Institute 2005; 84:1887-1896.

Iriti M, Di Maro A, Bernasconi S, Burlini N, Simonetti P, et al. Nutritional traits of bean (*Pisum sativum*) seeds from plants chronically exposed to ozone pollution.J Agric Food Chem.2009; 57:201–208.

Ito M, Deguchi Y, Matsumoto K, Kimura M, Onodera N, Yajima T. Influence of galactooligosacharides on the human fecal flora. Journal of Nutritional Science and Vitaminology.1993; 39:635-640.

Jin AL, Ozga JA, Lopes-Lutz D, Schieber A, Reinecke DM. Characterization of proanthocyanidins in pea (*Pisum sativum* L.), lentil (*Pisum sativum* L.), and faba bean (*Pisum sativum* L.) seeds.Food Res Int.2012; 46:528–535.

John S, Sophia JX, Yukio K, Sanja I, Daniel K. Isolation and characterization of lectins from kidney beans (*Pisum sativum*). Process Biochemistry.2007; 42:1436-1442.

Kahlon TS, Smith GE, Shao Q. In vitro binding of bile acids by kidney bean (*Pisum sativum*), black gram (Vigna mungo), bengal gram (Cicer arietinum) and moth bean (Phaseolus aconitifolins).Food Chem.2005; 90 :241–246.

Kalogeropoulos N, Chiou A, Ioannou M, Karathanos VT, Hassapidou M, Andrikopoulos NK. Nutritional evaluation and bioactive microconstituents (phytosterols, tocopherols, polyphenols, triterpenic acids) in cooked dry legumes usually consumed in the Mediterranean countries.Food Chem.2010; 121:682–690.

Kushi LH, Meyer KA, Jacobs DRJ. Cereals, legumes, and chronic disease risk reduction: evidence from epidemiologic studies.Am J Clin Nutr.1999; 70:451S–458S.

Lanza E, Hartman TJ, Albert PS, et al. High dry bean intake and reduced risk of advanced colorectal adenoma recurrence among participants in the Polyp Prevention Trial.J Nutr.2006; 136:1896–1903.

Lin LZ, Harnly JM, Pastor-Corrales MS, Luthria DL. The polyphenolic profiles of common bean (*Pisum sativum*.L.).Food Chem.2008; 107:399–410.

López-Amorós ML, Hernández T, Estrela I. Effect of germination on legume phenolic compounds and their antioxidant activity.J Food Comp Anal.2006; 19:277–283

Luthria DL, Pastor-Corrales MA. Phenolic acids content of fifteen dry edible bean (*Pisum sativum*.L.) varieties.J Food Comp Anal.2006; 19:205–211.

Madhujith T, Naczk M, Shahidi F. Antioxidant activity of common beans (*Pisum sativum*.L.).J Food Lipids.2004; 11:220–233.

Manach C, Scalbert A, Morand C, Remesy C, Jimenez L. Polyphenols: food sources and bioavailability.Am J Clin Nutr.2004; 79:727–747.

McBurney MI, Thompson LU. Effect of human faecal inoculum on in vitro fermentation variables.Brit J Nutr.1987; 58:233–243.

Nugent AP. Health properties of resistant starch.Nutr Bull.2005; 30:27–54.

Núñez CN. Natural antioxidants in health and diseases: A perspective. Journal of Environmental, Agricultural and Food Chemistry.2008; 7:3335-3342.

Nyström L, Schär A, Lampi AM. Steryl glycosides and acylated steryl glycosides in plant foods reflect unique sterol patterns.Eur J Lipid Sci Tech.2012; 114:656–669.

Oyeyinka SA, Singh S, Amonsou EO.Physicochemical properties of starches extracted from bambara groundnut landraces.Accepted manuscript: Starch Starke; 2016.

Panel, NCEPNE Third report of the National Cholesterol Education Program (NCEP) expert panel on detection, evaluation, and treatment of high blood cholesterol in adults (Adult Treatment Panel III) final report.Circulation.2002; 106:3143.

Parmar N, Virdi AS, Singh N, Kaur A, Bajaj R, Rana JC, Nautiyal CS. Evaluation of physicochemical, textural, mineral and protein characteristics of kidney bean grown at Himalayan region.Food Res Int.2014; 66:45–57.

Phillippy BQ. Inositol phosphates in food. Advances in Food and Nutrition Research. 2003; 45:1-60.

Pryor WA. Vitamin E and heart disease: basic science to clinical intervention trials. Free Radical Biology & Medicine. 2000; 28:41-164.

Ramírez-Jiménez AK, Reynoso-Camacho R, Tejero ME, León-Galván F, Loarca-Piña G. Potential role of bioactive compounds of.*Pisum sativum*.L. on lipid-lowering mechanisms. Food Res Int.2015; 76:92–104.

Rebello CJ, Greenway FL, Finley JW. A review of the nutritional value of legumes and their effects on obesity and its related co-morbidities.Obes Rev.2014; 15:392–407.

Remmelink M, Darro F, Decaestecker C, De Decker R, Bovin NV, Gebhart M., et al. In vitro influence of lectins and neoglycoconjugates on the growth of three human sarcoma cell lines. Journal of Cancer Research and Clinical Oncology.1990; 125:275-285.

Rocha-Guzmán NE, Herzog A, González-Laredo RF, Ibarra-Pérez FJ, Zambrano-Galván G, Gallegos-Infante JA. Antioxidant and antimutagenic activity of phenolic compounds in three different colour groups of common bean cultivars (*Pisum sativum*).Food Chem.2007; 103:521–527.

Ryan E, Galvin K, O'Connor TP, Maguire AR, O'Brien NM. Phytosterol, squalene, tocopherol content and fatty acid profile of selected seeds, grains, and legumes.Plant Food Hum Nutr.2007; 62:85–91.

Rycroft CE, Jones MR, Gibson GR &Rastall RA.A comparative in vitro evaluation of the fermentation properties of prebiotic oligosaccharides. Journal of Applied Microbiology. 2001; 91:878-887.

Sandberg AS. Bioavailability of minerals in legumes. British Journal of Nutrition 2002; 88:281-285.

Savage GP. Antinutritive factors in peas, In: Huisman J, Van der Poel TFB, Liner IE (eds.), Recent advances of research in antinutritional factors in legume seeds, Pudoc., Wageningen.1989; 4:342-350.

Segev A, Badani H, Kapulnik Y, Shomer I, Oren-Shamir M, Galili S. Determination of polyphenols, flavonoids, and antioxidant capacity in colored chickpea (Cicer arietinum.L.).J Food Sci.2010; 75: S115–S119.

Sharma S, Singh N, Virdi AS, Rana JC. Quality traits analysis and protein profiling of field pea (*Pisum sativum*) germplasm from Himalayan region.Food Chem.2015; 172:528–536.

Shevkani K, Singh N. Influence of kidney bean, field pea and amaranth protein isolates on the characteristics of starch-based gluten-free muffins.Int J Food Sci Technol.2014; 49:2237–2244.

Shevkani K, Singh N. Relationship between protein characteristics and film-forming properties of kidney bean, field pea and amaranth protein isolates.Int J Food Sci Technol.2015; 50:1033–1043.

Siah SD, Konczak I, Agboola S, Wood JA, Blanchard CL. In vitro investigations of the potential health benefits of Australian-grown faba beans (*Pisum sativum*.L.): chemopreventative capacity and inhibitory effects on the angiotensin-converting enzyme, α-glucosidase and lipase.Brit J Nutr.2012; 108: S123–S134.

Singh B, Singh JP, Kaur A, Singh N. Bioactive compounds in banana and their associated health benefits–a review.Food Chem.2016; 206:1–11.

Singh JP, Kaur A, Shevkani K, Singh N, Singh B. Physicochemical characterization of corn extrudates prepared with varying levels of beetroot (*Beta vulgaris*) at different extrusion temperatures.Int J Food Sci Technol.2016; 51:911–919

Singh JP, Kaur A, Shevkani K, Singh N. Influence of jambolan (*Pisum sativum*) and xanthan gum incorporation on the physicochemical, antioxidant and sensory properties of gluten-free eggless rice muffins.Int J Food Sci Technol.2015; 50:1190–1197.

Singh JP, Kaur A, Singh N, Nim L, Shevkani K, Kaur H, Arora DS. In vitro antioxidant and antimicrobial properties of jambolan (*Pisum sativum*) fruit polyphenols.LWT Food Sci Technol.2016; 65:1025–1030.

Terrés C, Navarro M, Martín-Lagos F, Giménez R, López H, López MC. Zinc levels in foods from southeastern Spain: Relationship to daily dietary intake. Food Additives and Contaminants.2001; 18:687-695.

Teuri U, Korpela R. Galacto-oligosaccharides relieve constipation in elderly people. Annals of Nutrition and Metabolism.1998; 42:319-27.

Tharanathan RN, Mahadevamma S. Grain legumes—a boon to human nutrition.Trends Food Sc Technol.2003; 14:507–518.

Tsuda T, Ohshima K, Kawakishi S, Osawa T. Antioxidative pigments isolated from the seed of.*Pisum sativum*.L.J Agric Food Chem.1994;42:248–251

Uddin M, Sarker M, Islam Z, Ferdosh S, Akanda M, Haque J, Yunus KB. Phytosterols and their extraction from various plant matrices using supercritical carbon dioxide: a review.J Sci Food Agric.2015; 95:1385–1394.

USDA (United Stetes Department of Agriculture). 2005. National Nutrient Database for Standard Reference, Release 18.

USDHHS (United States Department of Health and Human Services). US DA (Department of Agriculture). 2015. Dietary Guidelines for Americans (2015-2020). 8th Edition, Chapter 1.

Venn BJ, Mann JI. Cereal grains, legumes and diabetes.Eur J Clin Nutr.2004; 58:1443–1461.

Vergara-Castañeda HA, Guevara-González RG, Guevara-Olvera BL, Oomah BD, Loarca-Piña GF. Non-digestible fraction of beans (*Pisum sativum*.L.) modulates signalling pathway genes at an early stage of colon cancer in Sprague-Dawley rats.Brit J Nutr.2012; 108: S145–S154.

Vucenic I, Shamsuddin AM. Protection against cancer by dietary IP6 and inositol. Nutrition and Cancer.2006; 55:109-125.

Wang N, Toews R. Certain physicochemical and functional properties of fibre fractions from pulses.Food Res Int.2011; 44:2515–2523.

Wang YK, Zhang X, Chen GL, Yu J, Yang LQ, Gao YQ. Antioxidant property and their free, soluble conjugate and insoluble-bound phenolic contents in selected beans.J Funct Foods.2016; 24:359–372.

Weihrauch JL, Gardner JM. Sterol content of foods of plant origin.J Am Diet Assoc.1978; 73:39–44.

Xu B, Chang SK. Total phenolic, phenolic acid, anthocyanin, flavan-3-ol, and flavonol profiles and antioxidant properties of pinto and black beans (*Pisum sativum*.L.) as affected by thermal processing.J Agric Food Chem.2009; 57:4754–4764.

Xu BJ, Chang SKC.A comparative study on phenolic profiles and antioxidant activities of legumes as affected by extraction solvents. Journal of Food Science.2007; 72: S159-S166.

Yadav BS, Sharma A, Yadav RB. Effect of storage on resistant starch content and in vitro starch digestibility of some pressure-cooked cereals and legumes commonly used in India.Int J Food Sci Technol.2010; 45:2449–2455.

Yadav BS, Sharma A, Yadav RB. Resistant starch content of conventionally boiled and pressure-cooked cereals, legumes and tubers.J Food Sci Technol.2010; 47:84–88.

Yeh CT, Yen GC,Effects of phenolic acids on human phenolsulfotransferases in relation to their antioxidant activity. Journal of Agricultural and Food Chemistry.2003; 51: 1474-1479.

Zhang B, Deng Z, Ramdath DD, Tang Y, Chen PX, Liu R, Tsao R. Phenolic profiles of 20 Canadian lentil cultivars and their contribution to antioxidant activity and inhibitory effects on α-glucosidase and pancreatic lipase.Food Chem.2015; 172:862–872.

Zhang, J, Shi J, Ilic S, Jun X S, KakudaY.Biological properties and characterization of lectin from red kidney bean (*Pisum sativum*). Food Reviews International. 209; 25:12-27.

Zhao Y, Du SK, Wang H, Cai M. In vitro antioxidant activity of extracts from common legumes. Food Chem.2014; 152:462–466.

8

Biological Activation of Grains as Functional Food Ingredients

[1]Arashdeep Singh, [1]Savita Sharma and [1]Baljit Singh

[1]Department of Food Science and Technology, Punjab Agricultural University Ludhiana, Punjab

Introduction

Biological activation of grain or germination is a natural, biological processing techniques and traditional method that can be used to improve the nutritional, functional, and sensory properties of grains. During biological activation of grains, extensive phytochemical changes were reported in a variety of food grains, which reflects a dynamic and complex nutrient flux including remobilization, degradation and accumulation. During this process many hydrolytic enzymes were activated to hydrolyze the starches, proteins and, non-starch polysaccharides, which results in more oligosaccharides and free amino acids. Many nutrients and bioactive compounds reportedly increase and then decrease as they are utilized by the growing plant, which results in degradation of the high molecular weight polymers for generation of bio-functional substances. The biological activation of grains also realizes a significant growth of nutritional value by increasing the bioavailability of nutritional compounds, vitamins, bio-elements and other biologically active substances due to the partial hydrolysis of starch, proteins, hemicelluloses and even celluloses. Biological activation triggers the enzymatic activity in grain and dormant hydrolytic enzymes are activated, breaking down starch, fibers and proteins and leading to an increase in the amount of digestible compounds along with improvement of functional properties without any chemical modification. Also, the activity of some anti-nutritional factors (enzyme inhibitors, haemagglutinins, anti-vitamins etc.) decreases or disappears during activation, allowing a complete valorization of biological compounds in grains. During biological activation hydrolytic amylase enzymes, proteolytic enzymes, and catalase were activated in grains, which enhances the starch degradation, a complex biochemical process and thus lead to the conversion of insoluble granules to soluble starch and dextrin. Apart from positive shifting in the level of nutrition, biochemical activities which occur during activation also generate bioactive components of health-promoting activities through the enzymatic actions that are antioxidants in nature, such as ascorbic acid, tocopherols, tocotrienols and phenolic compounds, thus resulting in an increase of antioxidant activity. Other compounds generated during grain activation are γ-aminobutyric acid (GABA), inositols, ferulic acid, phytic acid, tocotrienols, γ-oryzanol, dietary fiber, zinc, potassium magnesium, phenolic compounds and prolylendopeptidase inhibitor. These bio-functional compounds generated during biological activation of grains posses many health promoting activities for human well being. Some of the health benefits exerted by biological activated grains are discussed below (Singh and Sharma 2017)

Antihyperlipidemic Effect

Hyperlipidemia in comprised of a heterogeneous group of disorders whose characteristic expression is an elevation in the plasma concentration of serum cholesterol. Many dietary factors have been shown to affect serum cholesterol levels. Whole grain consumption is associated with improved lipid status. Germinated brown rice has been demonstrated to have an effect of hypercholesterolemia. Studies reported that germinated brown rice in diet for 10 weeks showed decreased total cholesterol and low-density lipoprotein (LDL) cholesterol levels. Also the germinated brown rice water extract administration improved the serum lipid profiles of mice suffering from obesity. Both germinated brown rice and brown rice suppressed hypercholesterolemia induced by hepatoma growth by means of up-regulating cholesterol catabolism. The restorative effects of germinated brown rice on the hepatoma-induced decreases in HDL cholesterol levels were stronger than those of brown rice (Mamiya 2007). The changes in blood cholesterol could be modulated using germinated brown rice rather than non-germinated brown rice, suggesting that germinated brown rice is a more efficient functional diet food that affects high blood cholesterol. The excellent hypolipidemic action of germinated brown rice may be due to the high concentration of different biologically active components, such as GABA, dietary fibers, γ -oryzanol, or other antioxidants. Oxidative modification of LDL-cholesterol is considered to play a role in atherogenesis and coronary artery disease. Rice bran contains a variety of antioxidants, such as γ - oryzanol, tocopherols, tocotrienols, and phenolic compounds, which serve to protect against LDL oxidation (Wu et al., 2013).

Antihypertensive Effect

Germinated brown rice in diet helps in the prevention of blood pressure increase in hypertensive rats' patients and thus germinated brown rice can be used as part of a complementary therapeutic diet for hypertension. Spontaneously hypertensive rats fed a diet consisting of 40% pregerminated brown rice showed that pregerminated brown rice had a significant and strong antihypertensive effect in spontaneously hypertensive rats. The antihypertensive effect of germinated brown rice may be due to the complex actions of various components abundantly found in germinated brown rice, including GABA, dietary fiber, γ -oryzanol, and ferulic acid. Ferulic acid, whose content increases greatly after germination, had a beneficial effect in lowering the blood pressure in rats. Both the ethanol and lipid-soluble fractions of rice bran have a beneficial effect in lowering blood pressure and improving lipid profiles. Except for ferulic, other functional components, such as GABA, dietary fiber, and γ -oryzanol,may also contribute to the antihypertensive effect of rice bran (Mitsuo and Aya 2004).

Anti-cancer

The triterpene cycloartenol ferulate in rice bran counteracted the tumor-promoting activity in two-stage carcinogenesis in mouse skin. The brown rice and bran contain compounds that interfere with the proliferation or colony-forming ability of breast or colon cells and certain phenols contained in brown rice bran might be associated with its putative cancer chemopreventive properties. Arabinoxylan hemicellulose, an indigestible rice bran fiber macromolecule, decreased the exposure of rodents to carcinogenic xenobiotics by facilitating their intestinal excretion and also decreased hemicelluloses-inhibited 1, 2-dimethylhydrazine-induced intestinal carcinogenesis in rats. Studies also reported that rice germ had an inhibitory effect on azoxymethane-induced colon carcinogenesis, suggesting that brown rice possesses potential chemopreventive properties, and GABA might be associated with these properties. Tocopherols and tocotrienols, antioxidants in brown rice, have also been shown to exhibit tumor suppressor functions by many researchers. These results suggest that the consumption of brown rice and rice bran instead of white rice may be advantageous regarding cancer prevention. On this basis, germinated brown rice, which maintains the majority of bioactive constituents in rice bran and germ and even contains more bio-functional components than non-germinated brown rice, may be a good choice for a staple diet with respect to cancer prevention (Tatala et al., 2007).

Researchers also found that germinated brown rice extracts with enhanced levels of GABA inhibited leukemia cell proliferation and stimulated cancer cell apoptosis. GABA, which is abundant in germinated brown rice, was also reported to suppress tumor growth and germinated brown rice diet with high levels of GABA may prevent pulmonary adenocarcinoma in smokers. Germinated brown rice has potential to reduce mutations and improve human health because of the increases in insoluble, but hydrolysable, phenolic compounds during germination (Wu et al., 2013).

Antihyperglycaemia

The beneficial effects of germinated grains on fasting blood glucose levels, postprandial blood glucose levels, and insulin response in humans have been demonstrated in literature It was reported that both non-diabetic patients and hyperglycemic patients with uncontrolled diabetes who ate diets supplemented with germinated brown rice had lower postprandial blood glucose levels than those who ate white rice supplemented diets. The consumption of germinated brown rice as a staple food in patients with type 2 diabetes was useful in improving blood glucose and lipid levels. Germinated brown rice ameliorated the blood glucose, type-1 plasminogen-activator, and lipid peroxide concentrations in diabetic rats in comparison with diabetic rats fed white rice, suggesting that germinated rice intake may ameliorate diabetic vascular complications such as ischemic heart disease. The tocopherol and other antioxidants in germinated brown rice could also be associated with preventing the elevation of type-1 plasminogen activator and lipid peroxide concentrations. The postprandial glucose level was improved following the germinated bread intervention compared with the control bread. However, fasting insulin decreased with the control and insulin resistance effects were unclear, as well as significance between treatments. Whilst, it provides a preliminary indication that there may be a different, and potentially favourable, metabolic outcome when consuming a germinated wheat grain product, at least in healthy participants over a short time frame (Wu et al., 2013).

Lower Risk of Cardiovascular Disease (CVD)

As reported earlier for antihyperlipidemic and antihypertensive effects, germinated grains have beneficial effect of lowering blood pressure and LDL cholesterol, indicating that germinated brown rice has a role in preventing CVD. As suggested earlier for diabetes prevention, high postprandial blood glucose and low insulin concentrations are risk factors of CVD. A strong relationship was observed among a high carbohydrate diet, high blood glucose concentration, and the risk of CVD. The beneficial effects of germinated brown rice on fasting blood glucose level and postprandial blood glucose level and insulin response have already been demonstrated, which also indicates that germinated brown rice has a role in preventing CVD.

Reducing Risk of Alzheimer's Disease

Germinated brown rice facilitates spatial learning, in particular, it was surmised that germinated brown rice might prevent Alzheimer's disease associated with peptide-Aβ. Several mechanisms may contribute to the positive effects of germinated brown rice on learning and memory. Germinated brown rice contains a large amount of GABA, which plays a crucial role in memory processes. It is well known that GABA functions in animals as a major inhibitory neurotransmitter. Many neurological disorders, such as seizures, Parkinson's disease, stiff-man syndrome, and schizophrenia, have been shown to be related to alterations in GABA concentrations and glutamic acid decarboxylase (GAD) levels in the brain. In addition, germinated brown rice contains a much higher amount of ferulic acid than brown rice or white rice, which has been demonstrated to protect against Aβ-induced neurotoxicity and memory disturbance as well as learning and memory impairment induced by scopolamine or cycloheximide.

Other Health Effects of Germinated Grains

As dietary intervention in devolving countries germinated finger millet porridge could be used as a food with significantly greater soluble iron content and reduced phytates compared with control foods and this may have been adequate to correct iron deficiency and enhance the optimal for iron bioavailability. Also consumption of malted oats compared with non-malted oats significantly enhanced zinc absorption. Iron absorption was also higher, although this increase was not significant. Phytates measured by inositol phosphate content were reduced by 77% in the malted oats. This may explain the increased absorption of zinc and iron, as the germination processes initiate the activity of phytases that degrade phytates, thereby interfering with the complexing factors that bind minerals.

References

Mamiya T, Kise M, Morikawa K, Aoto H, Ukai M, and Noda Y (2007) Effects of pre-germinated brown rice on depression-like behavior in mice. Pharmacol Biochem Behav 86:62–67.

Mitsuo K and Aya M (2004). Pre-germinated brown rice as complementary therapeutic diet for hypertension. *Food Style* 8:54–57.

Singh A and Sharma S (2017). Bioactive components and functional properties of biologically activated cereal grains: A Bibliographic review. Critical Reviews in Food Sci Nutr, 57: 3051-3071.

Tatala E, Kagan V, Han D and Packer, L. (2007). Free radical recycling and intramembrane mobility in the antioxidant properties of alpha-tocopherol and alpha-tocotrienol. Free Radical Bio Medicine 10: 263–75.

Wu F, Yang N, Touré A, Jin Z and Xu X (2013). Germinated Brown Rice and Its Role in Human Health. Critical Reviews in Food Sci Nutr, 53:451-63.

9

Aromatic Spices Potent Sources of Natural Antioxidants

[1]Prachi Avinash and **[2]Arun Kumar**

[1]SMS (Nutrition) MLSU, Udaipur, Rajasthan
[2]Incharge, Department of Dairy and Food Chemistry, CDFST, Udaipur, Rajasthan

Introduction

India is considered as the "Home of Spices". They are a prized group of minor components and have been an integral ingredient of Indian diet. Since antiquity, spices form an important part of the Ayurvedic Pharmacopoeia (The Indian System of Medicine), with their widespread traditional use. With its wide climatic conditions and topographical features, India naturally possesses wide variety of medicinal flora. The country enjoys the distinction of being the largest producer and consumer of spices as well as the fastest growing spice market in the world.

Spices are potent plant foods that possess a wide range of nutrient bioactive compounds that contribute to improve health, while adding variety, colour, flavour and aroma to daily diet. The term spice or condiment applies to "such natural plant or vegetable products or mixtures thereof, in whole or ground form, as are used for imparting flavour, aroma and piquancy to and for seasoning food" (Sikri and Berwal, 2008).

Webster (2011) describes spices as 'Any of various aromatic vegetable productions as pepper, cinnamon, nutmeg, mace, allspice, ginger, cloves, etc., used in cookery to season and to flavour sauces, pickles, etc.; a vegetable condiment or relish, usually in the form of a powder; also, as condiments collectively'.

India alone contributes 25-30% of the total world trade in spices. It is interesting to note that nine spices namely pepper, ginger, clove, cinnamon, cassia, mace, nutmeg, pimento (allspice) and cardamom alone contributed as much as 90% of the total world trade. Of the 109 useful spices listed in the ISO, India produces 63 spices and exports approximately 10% of its produce to about 150 countries worldwide (Zachariah and Jayashree, 2011). Among the importing countries USA is the largest importer of spices. Several other countries like Australia, Britain, Canada, Russia, and some European countries also import spices (Panda, 2010).

Origin of Spices

The fame of Indian spices is older than the recorded history. Centuries before Greece and Rome had their birth sailing ships were carrying to Mesopotamia, Arabia and Egypt the Indian spices, perfumes and fine textiles. It was the lure of these spices that brought many seafarers to the shores of India. Long before the Christian era the Greek merchants thronged the markets of South India buying spices among other precious things.

It was in the year 1492, that Christopher Columbus discovered the New World. Five years later, four tiny ships sailed southward from the port of Lisbon Portugal under the guidance of Captain Vasco Da Gama. Like Columbus Vasco Da Gama too was searching for a new route to the spice lands of Asia. While Columbus failed to achieve that goal, Da Gama succeeded. In a two year 24,000 mile round trip, he took his ships around the continent of Africa to India and back to Lisbon. Only two of the four ships survived to reach their homeport. These two ships brought back a cargo of spices and other products worth 60 times the cost of the said voyage (Panda, 2010).

During these middle ages, a pound of ginger was worth a sheep, a pound of mace worth three sheep or half a cow. Pepper, the most valuable spice of all was counted out in individual peppercorns and a sack of pepper was said to be worth a man s life. Vasco Da Gama's successful voyage intensified an international power struggle for control over the spice trade. For three centuries afterward the nations of Western Europe, Portugal, Spain, France, Holland and Great Britain fought bloody sea was over the spice producing colonies. Writers acknowledge the fascinating history of spices as a story of adventure exploration conquest and fierce naval rivalry.

Classification of Spices

As per the agricultural classification of plants based on primary use, spices have been categorized as spice crops as plants grown for the production of aromatic materials or substances which are used as food flavouring or for other purposes because of their fragrance or preservative qualities. Spices are in solid or liquid forms (Bareja, 2010).

The taxonomic classification of spices is shown in table 1. A conventional classification of spices (table 2) based on degree of taste classifies spices into four categories, wiz hot spices, mild spices, aromatic spices and herbs and aromatic vegetables (Peter, 2001; Aggarwal, 2008).

Table 2.1: Taxonomic classification of spices

Angiospermae	Dicotyledoneae	Sympetalae			
				Solanaceae	chilli, paprika, red pepper
				Pedahaceae	sesame
			Campalunatae	Compositae	camomile, chicory, tarragon
		Archichlamydaeae	Piperales	Piperaceae	cubeba, long pepper, pepper
			Ranales	Myristicaceae	mace, nutmeg
				Lauraceae	bay leaf, cassia, cinnamon
				Magnoliaceae	star-anise
			Rhoeadales	Cruciferae	mustard, wasabi
			Myrtiflorae	Myrtaceae	allspice, clove
			Umbelliflorae	Umbel/rferae	anise, caraway, celery, chervil, coriander, cumin, dill, fennel, parsley
	Monocoty Doneae		Liliiflorae	Liliaceae	garlic, onion
				Iridaceae	saffron
			Scitamineae	Zingiberaceae	cardamom, ginger, turmeric
			Orchidales	Orchidaceae	vanilla

Table 2.2: Conventional classification of spices

Classes	Spices
Hot spices	Capsicum (chilies), Cayenne pepper, black and white peppers, ginger, mustard
Mild spices	Paprika
Aromatic spices	Allspice, cardamom, cassia, cinnamon, clove, cumin, dill, fennel, fenugreek, mace and nutmeg
Herbs	Basil, bay, dill leaves, marjoram, tarragon, thyme
Aromatic vegetables	Onion, garlic, shallot, celery

The Nutritive value of Spices

According to the Gopalan et al. (2007), spices are accessory foods mainly used for flavouring food preparations to improve their palatability. Some spices are however rich in iron, trace metals and potassium and also may provide ß-carotene and vitamin-C. Most spices contain high levels of tannin (turmeric) which interferes with absorption of iron. Phenols are the beneficiary antioxidant compounds found in spices in significant amounts. Halliwell (2002) and Aggarwal et al. (2002) revealed that owing to their antioxidant properties, various therapeutic effects have been assigned to spices and spice-derived ingredients.

Seed spices contain variable amounts of protein, fat, carbohydrate, fibre, minerals and vitamins. However, owing to the very small quantity used in the food, their contribution to the nutrient requirement is not significant. Proteins, carbohydrates, minerals and vitamins are thus relatively less important in delineating the quality of spices (Agarwal, 2001).

Spices are rich in essential oils which, in addition to be used in the food industry, are also used in cosmetic and medicinal preparations. The nutritive value of spices is insignificant, but their commercial value is high (Bareja, 2010).

Antioxidant benefits of spices

Extensive work is being conducted all over the world to demonstrate the anti-mutagenic and anti-carcinogenic potential of some of the commonly consumed spices. This renewed interest in natural medicines today is mainly due to the fact that many chronic diseases including cancer still remain difficult to cure. As such, attempts are being made to identify naturally occurring anticarcinogens, which may lead to new strategies for cancer prevention. In humans, phytoprotectants can have complementary and overlapping actions, including antioxidant, antimutagenic and anti-inflammatory effects, modulation of detoxification enzymes, and induction of apoptotic activity and so on. Thus, incorporation of spices in our regular diet may prove to be beneficial for our health. Nevertheless, the effect of spices in the context of total diet still remains to be evaluated. It is still not clear that mechanisms that appear to influence disease risk in animals, often fed high doses of these spices, can be extrapolated to humans consuming realistic amounts of these spices as part of their daily diet (Bhattacharjee and Sengupta, 2009). It is still unclear whether chronic consumption can lead to health risks (Hinneburg et al, 2006).

According to Tapsell et al. (2006) herbs and spices have a traditional history of use, with strong roles in cultural heritage, and in the appreciation of food and its links to health. Demonstrating the benefits of foods by scientific means remains a challenge, particularly when compared with standards applied for assessing pharmaceutical agents. Pharmaceuticals are small-molecular-weight compounds consumed in a purified and concentrated form. Food is eaten in combinations, in relatively large, unmeasured quantities under highly socialized conditions. The antioxidant properties of herbs and spices are of particular interest in view of the impact of oxidative modification of low density lipoprotein cholesterol in the development of atherosclerosis. A range of bioactive compounds in herbs and spices have been studied for anti-carcinogenic properties in animal research on the effects of herbs and spices on mental health should distinguish between cognitive decline associated with ageing and the acute effects of psychological and cognitive function. There is level I and II evidence for the effect of some herbal supplements on psychological and cognitive function. There is very limited scientific evidence for the effects of herbs and spices on type 2 diabetes mellitus.

A study by Jain et al. (2011) showed the antioxidant properties of 37 medicinal plants playing part significantly in postnatal recovery. The investigation supported the traditional use of these plants in postpartum care. It was quoted that some medicinal plants are potent antioxidants and may be efficient as preventive agents in many diseases.

Spices are an important group of agricultural commodities being used by many civilizations all over the world to aid flavour, taste and nutritional values in the food. In traditional medical systems, their ability to heal various physical, mental and emotional problems has widely been reported. With this view, HPLC analysis was performed to estimate phenolic acids in 21 spices (asafoetida, bishop's weed, black mustard, coriander, cinnamon, clove, curry leaf, cumin black, cumin, fennel, fenugreek, garlic, ginger, Indian cassia, Indian dill or dill large cardamom, onion, saffron, tamarind, true cardamom, yellow mustard) commonly used in India in different forms. Several parts of the spices, for instance, seeds, leaves, barks, rhizomes, latex, stigmas, floral buds and modified stems were used in the study. It was reported that spices are known to significantly contribute to the flavor, taste, and medicinal properties of food because of phenolics (Singh et al., 2004).

Spices used in Indian foods such as cloves (*Syzygium aromaticum*), licorice (*Glycyrrhiza glabra*), mace (aril of *Myristica fragans*), and greater cardamom (*Amomum subulatum*) were tested for their antioxidant properties *in vitro* by Yadav and Bhatnagar (2007). Cloves showed the highest DPPH radical scavenging activity, followed by licorice, mace and cardamom. FRAP values for cloves were also the highest, while other spices showed comparatively lesser FRAP values. The results show that the spices tested are strong antioxidants and may have beneficial effects on human health.

In a study by Souri et al. (2008) on the antioxidant activity and phenolic content of twenty four medicinal plant extracts, it was concluded that medicinal plants were promising sources of potential antioxidants and may be efficient as preventive agents in the pathogenesis of some diseases. Seven plant materials (cumin, fennel, lettuce, nutmeg, great plantain, common purslane and bishop's weed) showed high antioxidant activities.

Radical scavenging activity (RSA), antioxidant activity (AA), reducing power, total polyphenol (TP) and flavonoid con- tents (TFC) of oregano (*Oreganum vulgare* spp.), ajowan (*Trachyspermum ammi*) and Indian borage (*Plectranthus amboinicus*) extracts were evaluated by Khanum et al (2011). Oregano exhibited maximum radical scavenging activity (88.2%, 82.3%) for aqueous and ethanolic extracts at 50 ppm concentration respectively, followed by ajowan (86.9%, 68.4%) and Indian borage (30.5%, 30.4%). Extracts of oregano and ajowan showed better antioxidant activity and reducing power than that of Indian borage. Aqueous extracts of oregano had highest TP (Gallic acid equivalents) and TFC (Catechin equivalents) of 27.7% and 50.6% respectively compared to ajowan (6.7%, 24.4%) and Indian borage (4.2%, 5.5%). Synergistic studies showed that the addition of oregano extract appreciably enhanced the RSA of ajowan and Indian borage extracts even at 50 ppm concentration.

The relative levels of antioxidant activity, total flavonoid content, total phenolic content and reducing power of different organic and aqueous extracts with hexane sequentially extracted, dichloromethane, ethyl acetate, methanol and water of four different varieties of cardamom viz. Mysore, Malabar, Vazhukka and Guatemala were studied by Amma et al. (2010). Based on the results Malabar variety was identified as the best source of antioxidant compounds.

Skulas-Ray et al. (2011) found in an *in vivo* study that seasoning a meal high in fat with antioxidant-rich spice blend could enhance antioxidant defenses and reduce the body's negative responses. .Adding spices to the meal significantly increased the ferric reducing antioxidant power, such that postprandial increases following the spiced meal were 2-fold greater than that after the control meal.

In a study by Ninfali et al. (2005), twenty-seven vegetables, fifteen aromatic herbs and some spices consumed in Central Italy (the Marches region) were studied to reveal total phenolic, flavonoid and flavanol content as well as their antioxidant capacity measured by the oxygen radical absorbance capacity (ORAC) method. A comparison in terms of antioxidant capacity was made between different salads, as well as between salads to which aromatic herbs had been added. Among the spices tested, cumin and fresh ginger made the most significant contribution to the antioxidant capacity. It was concluded that spices and aromatized salts should be regarded as supplement seasonings capable of providing a marked increase in phenolic and antioxidant capacity. The addition of aromatic herbs to salads, at a percentage compatible with palatability, markedly increases their antioxidant capacity.

Muchuweti et al (2007) investigated aqueous methanolic extracts of 9 spices for their phenolic compounds composition and their antioxidant properties. The phenolic compound contents were determined by Folin Ciocalteau, tannin binding assay and HPLC. Antioxidant properties were determined by the reducing power assay, radical scavenging assay, and B-carotene linoleic model system. Among the spices investigated, cinnamon and oregano had the highest antioxidant activities of 61.76 and 58.28% respectively. It was concluded that spices had antioxidant activities better than ascorbic acid.

In a study by Hossain et al (2008) the antioxidant capacity of 30 spices used frequently in ready meals and a selection of key compounds from spices were investigated using ferric reducing antioxidant properties (FRAP), 2,2'-azinobis(3-ethylebenzothiaziline-6-sulfonate) (ABTS) and microsomal lipid peroxidation (MLP) assays. Antioxidant capacities of the spice extracts were compared to 5 popular synthetic antioxidants [buylated hydroxyanisole (BHA), butylated hydroxytoluene (BHT), tert-butylated hydroquinone (TBHQ), propyl gallate (PG) and octyl gallate (OG)]. Results showed that clove extracts had the highest antioxidant capacity as measured by FRAP, ABTS and MLP assays. Extracts from garlic powder were the lowest ranked of all the spices examined. Synthetic antioxidants were ranked in the following decreasing order of antioxidant activity PG > BHA > TBHQ > OG > BHT. Rosmarinic acid, a polyphenol commonly found in lamiaceae spices and eugenol from clove had higher antioxidant capacities than that of all synthetic antioxidants investigated. Antioxidant capacities of kaempferol from apiaceae spices, capsaicin from chilli, curcumin from turmeric, thymol from thyme and gingerol from ginger were also comparable to most of the synthetic antioxidants.

A study by Wang et al (2010) was conducted to examine the in vitro antioxidant activities of forty-five commonly used essential oils and their major components. The oils and major components were subjected to screening for their possible antioxidant activity by measuring the ABTS+ radical scavenging ability, reducing power and metal chelating activity. The ABTS+ radical scavenging ability and reducing power of cinnamon leaf and clove bud essential oils were the best two among these essential oils. Cinnamon leaf (96.45 ± 0.01%) and clove bud (96.33 ± 0.01%) essential oils showed the strongest ABTS+ radical scavenging ability. The EC50 values of cinnamon leaf and clove bud essential oils are 12 μg/mL and 10 μg/mL, respectively. Cinnamon leaf and clove bud essential oils showed reducing power of 119.42 ± 0.68% and 112.92 ± 0.87% relative to butylated hydroxyanisole (BHA), respectively.

References

Agarwal S. 2001. Seed Spices- An Introduction. Seed Spices: Production, Quality, Export. In Ed. By Agarwal S., Divakar Sastry E.V., Sharma R.K. Pointer Publishers, Jaipur. 2-18.

Aggarwal B.B., Ahmed N. and Muktar H. 2002. Spices as potent antioxidant with therapeutic potential. In Cadenas E., Packer L. (Ed.) Handbook of Antioxidants. Marcel Dekker, Inc., New York. 437-472.

Aggarwal R.A.K. 2008. Spices, condiments & Nutrition. In Spice crops of India. Ed. By Arya P.S. Kalyani Publishers, Noida. Ch. 4 25-26.

Amma K.P.A.P., Rani M.P. Sasidharan I. and Nisha V.N.P. 2010. Chemical composition, flavonoids-phenolic contents and radical scavenging activity of four major varieties of cardamom. International Journal of Biological and Medical Research. 1(3): 20-24.

Bareja B.G. 2010. Classification of agricultural crops. Crop review: http://www.cropsreview.com/support-files/agriculturalcrops-classification.pdf Retrieved 14/11/2012.

Bhattacharjee S. and Sengupta A. 2009. Spices in cancer prevention: An Overview. The International Journal of Nutrition and Wellness. 17(1).

Gopalan G., Sastri B.V.R. and Balasubramanian S.C. 2007. Nutritive Value of Indian Foods. National institute of Nutrition, ICMR, Hyderabad.

Halliwell B. 2002. Food derived antioxoidants: How to evaluate their importance food and In-vivo. In Handbook of antioxidants. Ed. By Cadenas E & Packer L, Marcel Dekker, Inc., New York. 1-45.

Hinneburg I., Dorman H.J.D. and Hiltunen R. 2006. Antioxidant activities of extracts from selected culinary herbs and spices. Food Chemistry. 97: 122-129.

Hussain A.I., Anwar F., Sherazi S.T.H. and Przybylski R. 2008. Chemical composition, antioxidant and antimicrobial activities of basil (Ocimum basilicum) essential oils depends on seasonal variation. Food Chemistry. 108: 986-995.

Jain N., Goyal S. and Ramawat K.G. 2011. Evaluation of antioxidant properties and total phenolic content of medicinal plants in diet therapy during post partum healthcare in Rajasthan. 4(3): 248-253.

Khanum H., Ramalakshmi K., Srinivas P. and Borse B.B. 2011. Synergistic antioxidant action of oregano, ajowan and Borage extracts. Food and Nutrition Science. 2: 387-392.

Muchuweti M., Kativu E., Mupure C.H., Chidewe C., Ndhlala A.R. and Benhura M.A.N. 2007. Phenolic composition and antioxidant properties of some spices. American Journal of Food Technology. 2(5): 414-420.

Ninfali P., Mea G., Giorgini S., Rocchi M. and Bacchiocca M. 2005. Antioxidant capacity of vegetables, spices and dressings relevant to nutrition. British Journal of Nutrition. 93: 257-266.

Panda H. 2010. Brief history of Spices. Handbook on Spices & Condiments (cultivation, processing and extraction). Paperback edition, Asia Pacific Business Press Inc.

Peter K.V. 2001. Handbook of herbs and spices, CRC Press, Woodhead Publishing Ltd, Cambridge, England. 1-12.

Sikri V. and Berwal J.S. 2008. Antioxidant effect of spices and their mixture in reference to cumin, garlic, ginger, mustard, red chilli and turmeric. Journal of dairying, foods and home science. 27(2): 120-124.

Singh U.P., Singh D.P., Maurya S., Maheshwari R., Singh M., Dubey R.S. and Singh R.B. 2004. Investigation on phenolics of some spices having pharmacotherapeutic properties. Journal of Herbal Pharmacology. 4(4): 27-42.

Skulas-Ray A.C., Kris-Etherton P.M., Tecter D.L., Chen C.Y., Vanden Heuvel J.P., West S.G. 2011. A high antioxidant spice blend attenuates post parandial insulin and triglycerides responses and increases some plasma measures of antioxidant activity in healthy overweight man. Journal of Nutrition. 141(8): 1451-1457.

Souri E., Amin G., Farsam H. and Barazandeh T.M. 2008. Screening of antioxidant activities and phenolic content of 24 medicinal plant extracts. DARU. 16(2): 83-87.

Tapsell L.C., Hemphill I., Cobiac L., Patch C.S., Sullivan D.R., Fenech M., Roodenrys S., Keogh J.B., Clifton P.M., Williams P.G., Fazio V.A. and Inge K.E. 2006. Health benefits of herbs and spices: the past, the present, the future. Medical Journal of Australia. 185(4): S4-24.

Wang H.F., Yih K.H. and Huang K.F. 2010. Comparative study of the antioxidant activity of forty-five commonly used essential oils and their potential active components. Journal of Food and Drug Analysis. 18 (1): 24-33.

Webster's Dictionary. Retrieved 12/11/2011 http://www.websters-online-dictionary.org/.

Yadav A.S. and Bhatnagar D. 2007. Free radical scavenging activity, metal chelation and antioxidant power of some of the Indian spices. Biofactors. 31(3-4): 219-227.

Zachariah T.J. and Jayashree E. 2011. Processing and value addition in spices for entrepreneurship development. Souvenir, National Seminar in Spice Processing and its impact on rural Economy. 17-18 Nov. 2011, Udaipur. 9.

10

Use of Functional Ingredients for Development of Healthy Meat Products

[1]A.R.Sen and [2]P.S. Girish

[1]ICAR-Mahatma Gandhi Integrated farming Research Institute, Motihari, Bihar
[2]ICAR-National Research Centre on Meat, Hyderabad, Telangana

Introduction

Food ingredients sector is indicating ample prospects for growth. The segment is an obligatory element for the growth of the food and beverage sector. These ingredients are also driving the development of the healthy functional foods industry to increase by 20% growth annually. Innovation in the food and beverage market, especially in the processed food domain, has created innumerable opportunities for new product development in the ingredients market as well. The segment is driven by the growing food processing industry and India is emerging as ingredients bowl of the future. Additionally, growing urbanization coupled with industrialization and changing lifestyles have led to growing demand for convenient food that offers health benefits, over and above the basic nutritional benefits of the foodstuff. There are nearly 10 to 15 broad classes of ingredients that include the very basic ingredients as well as certain value added ingredients that have niche applications. Key ingredients are preservatives, flavours, colours, hydrocolloids/stabilizers, emulsifiers, artificial sweeteners and value added health food ingredients such as omega-3-fatty acids prebiotics, probiotics etc. Until recently, there has been limited research into functional foods in meats, while functional foods in the dairy sector have been developed to a high degree. But meat is less allergenic than many other foods and much more research into the development and marketing of novel functional meat products is required. Such meat products would open up a new market for the meat industry. Nutraceutical and functional food industry is a very vibrant industry today with the claims and a clear mandate of health, wellness and prevention of diseases as one of its main goals.

Objectives for Using Functional Ingredients

The objective of including functional ingredients in meat products is not only to provide certain desirable sensory properties but also it is an attempt to entice health conscious consumers who are concerned with the content of fat, saturated fatty acids, and cholesterol and their association with cardiovascular diseases, certain types of cancers, obesity, and so forth.

Functional Modifications of Meat

A food can be regarded as functional, if it is satisfactorily demonstrated to beneficially affect beyond adequate nutritional effects, in a way that is relevant to either improved health or well-being and/or to a reduction in the risk of disease. As far as meat is concerned, the modifications to which it may be subjected to confer functional properties that are based on modifications to the feed an animal receives or on postmortem manipulation of the carcass or in the meat product formulation.

Dietary Enrichment of Specific Nutritional Compound

In recent years, farm animal dietary supplementation with vitamin E has increased the concentration of this compound in animal tissue to the extent that meat became a moderate vitamin E source. Supplementation with selenium, magnesium or iron has been employed to increase the concentration of these healthy elements in meat. Meat has started to be fortified with iron but this causes some problems such as rapid lipid oxidation and color deterioration. This problem was addressed by usage of iron encapsulation in liposomes before direct addition to meats. Meat and meat products are not perceived as good sources of both calcium and magnesium. However, the concentration in meat might be modulated by animal feed supplementation or direct addition of salt or brine to meats. Addition of magnesium to pigs' fodder improved the color of meat, reduced drip, exudative and cooking losses, decreased fat oxidation and increased acceptability scores. Addition of calcium and magnesium salts have also a positive effect on meat texture, especially calcium in activating calpain enzymes that lead to increased protein degradation what in turn increases tenderness, but also reduces emulsion stability, elasticity, cohesiveness and cooking yield.

Meat enriched with Vitamin E and selenium could be placed in functional food category and can substantially increase human Vitamin E and selenium status thus help in preventing various diseases especially cardiovascular diseases, cancer and age related diseases. In Korea, selenium-enriched pork "Selen Pork" was produced by feeding yeast bound selenium and sold as a functional food that can improve human health and nutrition. The addition of selenium to animal feeds increases lipid oxidation stability, delays microbial growth, and increases the firmness and chewiness of chicken meat. Pork meat enriched with organic selenium is characterized by lower drip loss, lower occurrence of pale meat and watery meat syndrome. Dietary supplementation of animals with vitamin C is widely studied. Frequently vitamin C is added to the feed in combination with vitamin E, because it protects vitamin E from oxidation processes via regeneration of α-tocopherol from tocopheryl radical. Direct addition of vitamin C to meat is difficult because it is not stable at the normal meat pH range. Zinc is an essential trace element for human health. The source of dietary zinc had an influence on feed intake and average daily weight gain. Inclusion in animal diet sources of organic zinc increased feed intake and body weight compared to inorganic zinc. At the same time decreased fat oxidation, reduced cooking loss, increased crude protein content and increased total antioxidant capacity in chicken meat were observed after feeding zinc.

Nutritional Manipulation

Fat content and fatty acid composition of meat may differ according to breed or genotype, the feeding background, and the muscle considered. Although beef usually has a P : S ratio around 0.1, its ratio n-6 : n-3 PUFA is particularly beneficial (around 2), especially from animals fed with grass containing high levels of PUFA n-3. The ratio of fat to lean in pig and cattle carcasses is influenced by the composition of diets and by the levels of feeding, particularly energy and protein intake. Dietary fatty acid composition is an extremely important part of the fatty acid profiles of monogastric animals (pigs, poultry) and a less important part in ruminants (cattle) if desirable combinations of fatty acids are to be obtained for human needs with less saturated and more mono- and polyunsaturated fatty acids. Interests in conjugated linoleic acid (CLA) have increased in the last decades as a result of its potential effects on human health-related benefits and animal production. It is possible to increase the content of CLA in meat from ruminant animals through the feeding diets with polyunsaturated fatty acid-rich diet. The eggs rather than the poultry meat industry has progressed faster in terms of production and marketing of omega 3 fatty acids enriched products.

It is generally accepted that dietary CLA can improve the body composition through reducing fat deposition and backfat thickness. Dietary CLA not only reduced fat deposition but also altered the fatty acid composition of tissue lipids. Generally, ruminant meat has greater concentration of CLA than that from non-ruminants. CLA can be naturally synthesized in the rumen of ruminant animals by bacteria *Butyrivibriofibrisolvens* via the Δ-9-desaturase of trans 11 octadecanoic acid pathway. Therefore, it is possible to increase the content of CLA in meat from ruminant animals through the feeding diets with polyunsaturated fatty acid-rich diet.

Meat Product Formulation

Meat products are certainly main source of food proteins with high biological value in various countries. Raw, partially cooked, fully cooked, ready-to-eat, fermented, dried, injected, marinated and dry-cured products all derive distinctive properties from use of non-meat ingredients. Change in the use of the basic functional non-meat ingredients is virtually guaranteed to require an alternative ingredient or multiple ingredients, and/or process adjustments to achieve the same or similar finished product properties. Modification of meat formulation process also makes it possible to use traditional ingredients and other ingredients specifically designed with certain attributes that confer health-promoting properties.

Modification of Fat, Fatty Acids and Cholesterol

Fat reduction is usually based on two main criteria, utilization of leaner meat raw materials and the reduction of fat by adding water and other ingredients like carbohydrate based fat replacers or gums or protein based fat replacers with little or no calorific content. The second procedure consists of replacing part of the animal fat normally present in the product with another more suited to human needs, i.e. with less saturated fatty acids and more monounsaturated (oleic) or polyunsaturated acids, and with no cholesterol. Although simple fat replacement does not reduce the caloric content, it does greatly improve the nutritional qualities of the product. Both fish oils (omega-3 polyunsaturated oil) and vegetable oils (partially hydrogenated from corn, cottonseed, palm, peanut and soybean, peanut, high-oleic acid sunflower, cottonseed, olive) have been used for this purpose in products such as patties and sausages.

Addition of vegetable oils

Olive oil is a monounsaturated vegetable oil. It has a high biological value, and its consumption is related to a decreased risk of heart disease and breast cancer. Vegetable oils have also been used as partial substitutes of pork back fat in low-fat frankfurters and other types of cooked meat product giving rise to products with more adequate fatty acid profiles and cholesterol levels. An alternative to using this vegetable oil, which has a high unsaturated fatty acid content and is liquid at room temperature, is to use inter esterified vegetable oils (IVOs). These oils can be used as a fat replacer to modify the fatty acid composition of frankfurters and Turkish type salami without any detrimental changes in sensory characteristics. Linseed oil has greater potential in this area. The substitution of pork back fat with linseed oil in the manufacture of dry-fermented sausages decreased the n-6:n-3 ratio (from 14.1 to 2.1) as a consequence of the increase in α-linolenic acid.

Addition of Healthy Fatty Acids

It is known that omega-3 fatty acids (n-3 FA) have potential in the prevention and treatment of cardiovascular diseases, some autoimmune disorders, diabetes, and some types of cancer aside from their important role in neuronal development.Omega-3 polyunsaturated fatty acids, especially the long chain (LC) fatty acids, eicosapentaenoic acid (EPA) and docosahexaenoic acid (DHA), found in fatty fish and fish oil, are well recognized for their protective effects on cardiovascular disease, inflammatory diseases (such as rheumatoid arthritis), their role in brain health (improve cognitive function and mood and possibly protect against dementia) and various other biological functions. Oils in the form of n-3 polyunsaturated fatty acids occur mainly in cold water fish, whereas n-6 polyunsaturated fatty acids come mainly from plants and saturated fatty acids from animal sources. Vegetable oils including linseed oil and rapeseed oil could also increase ω3 fatty acid content in the form of linolenic acid, which could be used to synthesize long chain ω3 PUFA.

Addition of Sodium Chloride, Phosphates and Curing Ingredients

Salt (sodium chloride) is clearly one of the most important functional non-meat ingredients in processed meats, and probably the most common in terms of both quantity and frequency of use. The importance of salt is widely recognized due to the multifunction role of this ingredient in processed meats, some of which are indispensable. Because salt dissociates into sodium and chloride ions in the water phase of meat mixtures, its functions are largely the result of the dissociated ions. Chloride ions provide much of the functionality of salt for water binding and yields, while sodium ion provides the flavour expected of products with added salt. Chloride ions in a cured meat product may also play a small but potentially significant role in cured color development because chloride can accelerate the conversion of nitrite to nitric oxide during the cure development phase prior to cooking of cured meat product. Due to the role of sodium in the development of hypertension in sodium-sensitive individuals, public health and regulatory authorities have recommended a reduced dietary intake of sodium choride. Since salt contributes to water and fat binding in meat products, its reduction has an adverse effect on these parameters increasing cooking loss and weakening the texture. Partial replacement of NaCl by calcium ascorbate seems to be a viable way of decreasing sodium in dry-fermented sausages.

Another multifunctional ingredient is the phosphates which are widely used in meat processing. While the phosphates are yet another multifunctional ingredient, the phosphates are different from the other multifunctional ingredients in processed meats as several different forms of the phosphates are used. The primary function of the phosphates is to increase water retention by meat proteins to improve product yields and improve product tenderness and juiciness. Phosphates have been shown to be significant contributors to antioxidant activity and flavor development in processed meats, particularly in uncured products where nitrite is not included.

Nitrite, similar to sodium chloride, is a multifunctional ingredient but nitrite is also much more potent than sodium chloride in its effects on meat products. While nitrite will react with meat components quite readily, it is often slower than desired in high-volume production environments. Consequently, curing accelerators have been added as another ingredient to the meat processing because the reduction of nitrite to nitric oxide is an absolutely essential step in the development of pinkish cured meat color and probably most of the other cured meat properties as well. The ingredients most commonly used as cure accelerator like sodium ascorbate, sodium erythorbate, ascorbic acid and erythorbic acid which react directly with nitrite to produce nitric oxide and nitroso haemochromogen for producing cured meat colour.

Addition of Dietary Fibres

Dietary fibre is the remnant of the edible parts of plants or analogous carbohydrates that are resistant to digestion and absorption in the human small intestine with complete or partial fermentation in the large intestine. Dietary fibre is a key functional ingredient widely used nowadays while developing designed foods due to its significance in health promotion and technological impact. Dietary fibre is among the food ingredients, now called nutraceuticals, which are shown to have the potential, if consumed in adequate amounts, to improve human health. Fibre is one of the valuable components that can be incorporated in meat products from health point of view.

Various fiber sources like oat, rice, sugar beet, soy, pea, psyllium etc. have been tried in formulation of some meat products such as patties and sausages for development of nutritionally balanced diet (Table 1). Fiber is also being used as a fat replacer in manufacturing of various meat products. Other than the nutritional properties, dietary fiber is now used in meat products for technological up gradation like improvement in cooking yields, rheological properties, reducing formulation costs and enhancing the palatability and texture of meat products.

Table 1: Various fibre sources utilized for formulation of fibre rich meat products

Type/source of fiber	Type of meat product
High soluble fibre sources	Sausages
Fructo- oligosaccherides	Sausages
Inulin	Mortadella
	Veal meatballs
Carrageenan	Mutton kofta
Konjac flour	Frankfurters
Carboxy methyl cellulose	Sausages
Legumes (hulls and flours)	Beef patties
	Meat balls
	Buffalo meat burgers
	Chicken nuggets
	Camel meat patties
	Beef burgers
	Goat meat patties
	Chicken loves
	Buffalo meat patties
Cereals and bran	
Oat bran	Meat balls
	Pork sausages
Oat flour	Beef patties
	Mutton kofta
	Chicken kofta
Wheat bran	Meat balls
	Beef patties
	Chicken patties
Rice bran	Pork meatball
	Meat batter, sausages and emulsions
	Frank furters
Rye bran	Meat balls
Barley	Bologna
	Pork patties
Ragi millet	Chicken patties
Fruits, vegetables and their by- products	
Apple pulp	Chicken nuggets
	Fermented sausages
Lemon albedo	Sausages
Orange	Fermented sausages
	Bologna
Kinnow	Goat meat patties
Peach dietary fiber	Frank furters
	Fermented sausages

Type/source of fiber	Type of meat product
Carrot dietary fiber	Sobrassada
	Chicken nuggets
Sugarbeetfiber	Frank furters
	Turkish type salami
Oyster mushroom	Chicken patties

Source: Mehta et al., 2015

Inherent functional properties of fiber have a profound effect on physicochemical properties. Fiber addition in meat results in decreasing cooking loss, alteration in pH and an increase in emulsion stability. Oat bran or oat fibre appears to be a suitable fat replacement in ground beef and pork sausage products due to its ability to retain water and emulate particle definition in ground meat in terms of both colour and texture. Fiber is suitably used in cooked meat products to increase the water holding capacity and resulting juiciness and increasing yield of product. Fiber addition in meat products can result in a change in overall composition of the final products. The moisture content and carbohydrate level increased with a subsequent increasing level of water and oat bran in frankfurter. Another important source of fiber is fruits, which can also be obtained as by-products of plant food processing. Citrus byproducts (lemon albedo and orange fiber powder) have been added, at different concentrations, to cooked and dry-cured sausages with excellent results. With regard to products quality, fruit fibres have been found more desirable compared to fibres obtained from cereals. Various sources markedly enhance the dietary fibre content in meat products and making them more functional as well as healthier.

A. Addition of Plant Proteins

Protein ingredients are the main vegetable component used in manufacturing meat products, for technological purposes, cutting costs as well as for nutritional benefits, reducing the cholesterol level, increasing the protein components and improving the amino acid profile. Soy protein has been used as functional ingredients in various meat products such as low fat sausage. Soya and sunflower proteins, wheat and maize derivatives, and flours from cottonseed and oats, some of which contain health-enhancing substances, have all been used as fat replacements. Soy proteins are widely used in meat products in the forms of soyflour, and soy protein concentrate and isolate to improve water and fat binding ability, enhance emulsion stability, improve nutritional content and increase yields. However, soy flour produce some beany flavor and soy protein concentrates and isolates provided some undesirable palatability in meat products added with soy proteins. To overcome these disadvantages, dried soy tofu powder was added in frankfurters and pork sausage patties. Incorporation of tofu powder resulted in lower fat and higher protein and moisture content, but did not affect sensory parameters in lean pork sausages. Wheat proteins could be a great additive due to their ability to form viscoelastic mass of gluten through the interaction with water.

Use of Natural Antioxidants

A number of strategies are being used to enhance antioxidant activity in meat systems by adding natural antioxidants (flavonoids/phenolic compounds) and to reduce the formation of oxidation products causing ageing, cancer and cardiovascular disease.Lipid oxidation is one of the causes for the deterioration of meat and their derivatives determine the onset of a large number of undesirable changes in flavor, texture, and nutritional value. Due to health concern of synthetic antioxidants natural antioxidants extracted from plants such as rosemary, sage, tea, soybean, citrus peel, sesame seed, olives, carob pod, and grapes can be used as alternatives to the synthetic antioxidants because of their equivalent or greater effect on the inhibition of lipid oxidation (Table 2). Phytochemicals such as phenolics, anthocyanins and other flavonoids prevent the occurrence of certain chronic diseases. Flavonoids are unique novel molecules possessing strong antioxidant capacity. Meat and meat products become rich in antioxidants by adding plant phytochemicals and increases the value of meat. Compounds from herbs and spices contain many phytochemicals which are potential sources of natural antioxidants including phenolic diterpenes, flavonoids, tannins and phenolic acids. These compounds have antioxidant, anti-inflammatory and anticancer activities. Antioxidants vary widely in chemical structure and have varied mechanisms of action. The key mechanism is their reaction with free radicals to form relatively stable inactive products. Thus, antioxidants delay lipid oxidation by scavenging free radicals which are generated in the initiation phase, propagation phase, or during the breakdown of the hydroperoxides.

Rosemary extract contains high levels of phenolic compounds leading to its great antioxidant activity. Phenolic compounds are capable of regenerating endogenous tocopherol in the phospholipid bilayer of lipoprotein. Catechins is a predominant group of polyphenols present in green tea leaves composed of four compounds epicatechin, epicatechingallate, epigallocatechin, and epigallocatechin gallate. These tea compounds promote health by preventing lipid oxidation and providing antibacterial, anticarcinogenic and antiviral ability. Grape seed extracts contain a number of polyphenols including procyanidins and proanthocyanidins are powerful free radical scavengers. Seaweeds are considered to be a rich source of antioxidants. Recently, the potential antioxidant compounds were identified as some pigments (fucoxanthin, astaxanthin, carotenoid etc.) and polyphenols (phenolic acid, flavonoid, tannins etc.). Those compounds are widely distributed in plants or seaweeds and they exhibit higher antioxidant activities. Seaweeds are noted to contain not only labile antioxidants (i.e., ascorbate, glutathione) when fresh but also, more stable molecules, such as carotenoids, mycosporine like amino acids and a variety of polyphenols like catechins and phlorotannins.

Table 2: Natural antioxidants used in production of functional meat products

Source	Main active compound	Mode of action	Meat product
Red grapes, gooseberry and tomato	Polyphenolic compounds	Radical highly oxidizable polyunsaturated fatty acids	Restructured irradiated pork loins.
Rosemary-tocopherol	Tocopherol and volatile acid compounds	Protects highly oxidizable polyunsaturated fatty acids	Restructured irradiated pork loins
Walnuts	High-biological-value proteins, vegetable fiber, polyunsaturated (linoleic and lino-lenic) fatty acids (especially g-tocopherol), and other antioxidants (phytosterols and polyphenols),	Radical scavenger	Restructured beef steak
Rosemary extract, chitosan and carnosine	Phenolic diterpenes such as carnosic acid, carnosol, rosmanol, rosmariquinone and rosmaridipheno	Act as metal chelators and singlet oxygen quenchers	Beef patties w
Noni (Morinda citrifolia)	Antraquinones	Radical scavenger	Chicken breast hamburger
Grape antioxidant dietary fiber (GADF)	Dietary fiber and phenolics antioxidants such as pehnolics acids, anthocyanidins, proanthocyanidins, catechins and other flavonoides is	Radical scavenger	Chicken breast hamburger
Olive waste extract	Tyrosol related compounds and derivatives of benzoic and cinnamic acids	Radical scavenger act as metal chelators and singlet oxygen quenchers	Lamb meat patties
Aloe Vera	Anthraquinone, phenolic compounds	Radical scavenger	Goat meat nuggets

Source: Mireles-Arriaga et al. 2017

Bioactive Peptides in Meat Products

Being a major source of high quality proteins and important source of essential amino acids, muscle food presents one of the most investigated sources for isolation of bioactive peptides in recent number of years. These can be used as potential functional ingredient for using in different food model system. Although most peptides derived from animal sources are generated from milk and milk-based products proteins, peptides also were isolated from meat proteins, eggs, bovine blood, collagen, gelatin, various fish species including tuna, sardine, herring, salmon, bonito and from marine organisms. Bioactive peptides are considered the new generation of biologically active regulators that not only prevent the mechanism of oxidation and microbial degradation in foods but also enhanced the treatment of various diseases and disorders, thus increasing quality of life. In general, bioactive peptides usually contain 2–20 amino acid sequences with molecular masses of less than 6 kDa, and their activities are based on the inherent composition and sequence of amino acids.

A numerous methods have been utilized to release bioactive peptides from muscle proteins and enzymatic hydrolysis of protein is the most widely used technique. For example, a number of bioactive peptides have been isolated from meat using digestive enzymes such as pepsin, trypsin and chymotrypsin. Use of acid hydrolysis in order to release some peptides is economic, relatively simple to perform, but at the same time difficult to control and can damage certain amino acids. Methods based on enzymatic hydrolysis have an advantage because they are more predictable with respect to the end products, and the process conditions can be controlled. Because of that, these are commonly used methods for peptide production in laboratories and industry. Although bacterial fermentation presents valuable method for isolation of bioactive peptides from milk proteins, it wasn't successful in obtaining peptides from meat and fish source, probably because of poor proteolytic activity of the *Lactobacillus* spp. used in meat fermentations.

Functional Ingredients as Starter Culture

A probiotic is known as a culture of living microorganisms which are mainly lactic acid bacteria or *Bifidobacteria*. It can beneficially affect human health when it is ingested at certain levels by preventing the growth of harmful bacteria via competitive exclusion and by generating organic acids and antimicrobial compounds in the colon. Meat products, processed by fermentation without heating have been considered an excellent vehicle for probiotics. Lactic acid bacteria (LAB) are essential agents during meat fermentation improving hygienic and sensory qualities of the final product. The use of probiotics as functional ingredient seems more promising in raw fermented meat products like salami as they are made with raw meat and consumed without prior heating, which would kill the probiotic bacteria. However, the incorporation of probiotic bacteria to these products also represents a technological challenge because of the known sensitivity of probiotic to curing salts, spices and other ingredients used in the formulation of the fermented sausages. Furthermore, this addition requires the use of microorganisms that are resistant to the fermentation process and that remain in a minimal viable number of cells to survive the stomach pH and exert beneficial effects in the intestines. The desirable probiotics should have following properties: resistance to acid and bile toxicity; adherence to human intestine cells; colonization in human guts; antagonism against pathogenic bacteria; production of antimicrobial substances; and immune modulation properties.

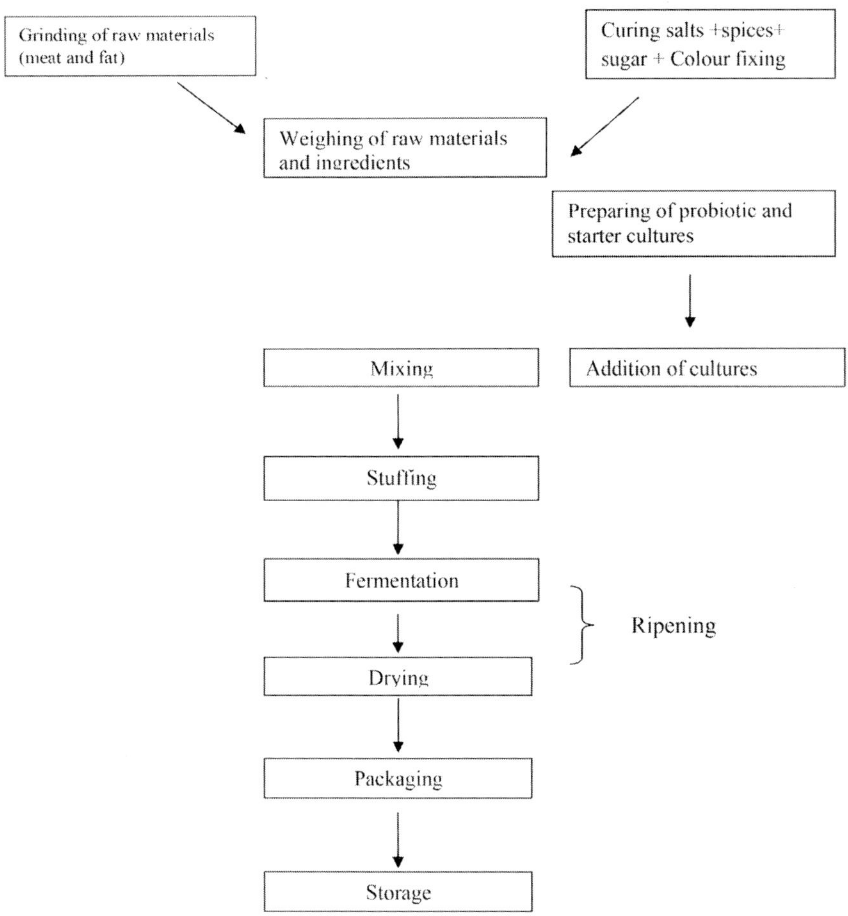

Fig. 1: Basic flowchart of the processing of fermented dry sausage with the of probiotic cultures

Source: Leory et al., 2006

Fermentation of meat causes number of physical, biochemical and microbial changes, which eventually result in functional characteristics of the products (Fig. 1). Those changes include acidification (carbohydrate catabolism), solubilization and geleation of myofibrilla and srcoplasmic proteins, degradation of proteins and lipids, reduction of nitrate into nitrite, formation of nitrosomyoglobin and dehydration. Degradation of proteins during the fermentation process is one of the key factors involved in the improvement of functional value of meat products producing bioactive peptides. Bacteriocins are the peptides produced by lactic acid bacteria with antibacterial properties. These peptides can reduce or inhibit the growth of other Gram-positive bacteria. Probiotic bacteria and their metabolized products have various functions including modulation of intestinal flora, prevention of diarrhea, improvement of constipation, prevention and treatment of food allergies, reduction of cancer risk, lowering of plasma cholesterol etc.

Conclusion

Functional foods provide good opportunity for the meat industry, in order to improve the quality of meat, and create meat products with health beneficial properties. Meat and meat products are excellent foods for delivering bioactive compounds without changing dietary habits. Some bioactive compounds from fruits and vegetables (walnuts, tomatoes, soy protein isolate, red beetroot juice and vegetable oils) appear to play an important role in the prevention of specific diseases like cardiovascular diseases, cancers and diabetes mellitus. Meat and meat products are considered to be the best foods for delivering bioactive compounds because the meat is frequently consumed and has a good nutritional value which can be modified easily according to specific needs. Many strategies have been devised to develop healthier meat products. The producers of these meat based functional foods claim that these products have such bioactive compounds in optimized concentration which can promote health and decrease disease risk factors. However, there are fewer studies to evaluate the effect of these compounds on health. We need more advancement in the field of human intervention studies. Meat products are not a traditional matrix which include functional ingredients, their incorporation imply significant technological challenges and multidisciplinary approaches are needed for using them in different meat products.

References

Bhat, Z.F. and Bhat, H. (2015). Fibre based functional meat products. Asian Journal of Food and Agro-Industry. 4:261-271.

Decker, E.A. and Park, Y. (2010). Healthier meat products as functional foods. Meat Science. 86: 49-55.

Fernandez-Gines, J.M., Fernandez-Lopez, J., Sayas-Barbera, E. and Perez-Alvarez, J.A. (2005). Meat products as functional foods: A review. Journal of Food Science. 70: R37-R43.

Fernandez-Lopez, J., Fernandez-Gines, J.M., Alleson-Carbonell, L., Sendlara, E., Sayas-Barbera, E. and Perez-Alvarez, J.A. (2004). Application of functional citrus by-products to meat products. Trends in Food Science and Technology. 15:176-185.

Jimenez-Colmenero, F., Carballo, J. and Cofrades, S. (2001). Healthier meat and meat products: their role as functional foods. Meat Science. 59: 5-13.

Kumar, P., Kumar, S., Tripathi, M.K., Mehta, N., Rajan, R., Bhat, Z.F. and Singh, P.K. (2013). Flavonoids in the development of functional meat products: A Review. Vet World. 6:573-578.

Kumar, Y., Yadav, D.N., Ahmad, T. and Kairam, N. (2015). Recent Trends in the Use of Natural Antioxidants for Meat and Meat Products. Comprehensive reviews in food science and food safety. 14: 796-812.

Leroy F, Verluyten J, and Vuyst L. (2006). Functional meat starter cultures for improved sausage fermentation. International Journal of Food Microbiology. 106:270-285.

Mehta, N.,Ahlawat, S.S., Sharma, D.P. and Dabur, R.S. (2015). Novel trends in development of dietary fibre rich meat products-a critical review. Journal of Food Science and Technology. 52:633-647.

Mireles-Arriaga, A.I., Ruiz-Nieto, J.E., Hernandez-Ruiz, J., Juarez-Abraham, M.R., Sanzon-Gomez, D. and Mendoza-Carrillo, M. (2017). Functional restructured meat: Application of ingredients derived from plants. Vitae. 24:196-204.

Naveen Kumar, T.J., Nayar, R. and Rajagopal, K. (2015). Technological and health impacts of meat based functional products-A review. International Journal of Science and Research. 4: 475-480.

Pogorzelska-Nonicka, E., Atanasov, A.G., Horbanczuk, J. and Wierzbicka, A. (2018). Bioactive compounds in functional meat products. Molecules. 23: 307-326.

Sebranek, J.G. (2015). An overview of functional non-meat ingredients in meat processing: the current toolbox. 68th Annual Reciprocal Meat Conference.42-46.

Sen, A.R. and Mandal, P.K. (2017). Use of natural antioxidants in muscle foods and their benefits in human health: An overview. International Journal of Meat Science. 7:1-5.

Verma, A.K. and Banerjee, R. (2010). Dietrayfibre as functional ingredient in meat products: a novel approach for healthy living-a review. Journal of Food Science and Technology. 47: 247-257.

Zhang, W., Xiao, S., Samarweera, H., Lee, E.J. and Ahn, D.U. (2010). Improving functional value of meat products. Meat Science. 86:15-31.

11

Functional Foods Ingredients from Fish and Fish Products

Sahar Masud

Associate Professor, Department of Fisheries, Sher-e-Kashmir University of Agricultural Sciences and Technology, Jammu, Jammu & Kashmir

Introduction

Increasing knowledge regarding the impact of diet on human health along with state-of-the-art technologies has led to significant nutritional discoveries, product innovations, and mass production on an unprecedented scale. In particular, naturally occurring bioactive extracts or single compounds thereof, that are believed to benefit human health, have spawned an important and dynamic new area of research resulting in substantial advances in nutritional knowledge. There is also growing awareness that dietary source and form of food may affect overall health. Suitably, the role of food as an agent for improving health has been recognized, initiating the development of new classes of food, known as functional foods (Honkanen 2009).

The concept of functional foods is to improve the general conditions of the body and decrease the risk of illness and disease. That is to say, bioactive compounds present as natural constituents or as fortificants in food having the potential to provide health benefits beyond the basic nutritional value of the product. Marine-derived nutrients and other marine bioactive components have excellent potential as functional food ingredients as they possess advantageous physiological effects, with medicinal characteristics and added health benefits such as anticancer or anti-inflammatory activity.

The marine world, due to its phenomenal biodiversity, is a rich natural resource of many biologically active compounds such as polyunsaturated fatty acids (PUFAs), sterols, proteins, polysaccharides, antioxidants and pigments. Many marine organisms live in complex habitats exposed to extreme conditions and, in adapting to new environmental surroundings, they produce a wide variety of secondary (biologically active) metabolites which cannot be found in other organisms. Moreover, considering its great taxonomic diversity, investigations related to the search of new bioactive compounds from the marine environment can be seen as an almost unlimited field.

Marine-based bioactive food ingredients can be derived from a vast array of sources, including marine plants, microorganisms, and sponges, all of which contain their own unique set of biomolecules. However, proving that these naturally occurring bioactive substances have a defined health benefit poses a dilemma in nutritional research as investigating preventive activity can be difficult when effect is only moderate. This means that the effect of the compounds on the human body may be very small over relatively short periods but could contribute significantly to health when they are consumed throughout life as part of the daily diet. Therefore, to facilitate discussion of this issue, the following review examines the existing scientific knowledge which demonstrates the suitability of marine-derived bioactive compounds as functional food ingredients for the prevention and treatment of chronic diseases (Rasmussen and Morrissey 2007).

Sources of Marine Functional Food Ingredients

Macroalgae

Macroalgae are a source of biologically active phytochemicals, which include carotenoids, phycobilins, fatty acids, polysaccharides, vitamins, sterols, tocopherol and phycocyanins among others. Many of these compounds are known to possess biological activity and hence have potential beneficial use in healthcare. However, the chemical and nutritional composition of seaweeds depends on many factors, including species, geographical origin or area of cultivation, seasonal, environmental, and physiological variations, time of harvest, water temperature, and processing methods (Bocanegra et al. 2009).

Proteins, Peptides and Amino Acids

The protein content of macroalgae varies greatly from phylum to phylum. Generally, the protein fraction of brown seaweeds is low (3–15% of dry weight) compared with that of the green or red seaweeds (10–47% of dry weight). The protein in macroalgae contains all essential amino acids; however, variations in their concentrations are known to occur. Leucine, valine, and methionine are abundant essential amino acids of.*Palmaria palmata*.and their average levels are close to those generally reported for ovalbumin. On the other hand, isoleucine and threonine concentrations are similar to those recorded for legume proteins.

Leucine, phenylalanine and valine are the major essential amino acids of.*Ulva rigida*, while levels of histidine, which is an essential amino acid in children, are similar to those found in legumes and eggs. Bioactive peptides usually contain 3–20 amino acid residues and their activities are based on their amino acid composition and sequence. These peptides are reported to be involved in various biological functions such as antihypertension, immunomodulatory, antithrombotic, antioxidant, anticancer and antimicrobial activities, in addition to nutrient utilisation (Fleurence 1999).

Among the algal proteins, it is worth noting the occurrence of protein-pigment complexes called phycobiliproteins, some of which are currently used as fluorescent markers in the fields of clinical diagnosis and biotechnological applications. Recent studies have shown that phycobiliproteins, which generally make up 1–10% of dry weight of algal biomass, impart antioxidant properties which could be beneficial in the prevention or treatment of several diseases. Moreover, in some countries, phycobiliproteins are utilised as natural food colourings in products such as chewing gums, dairy products, jellies and ice sherbets (Aneiros 2004).

Fatty Acids

The lipid content of macroalgae represents only 1–5%, thus its contribution as a food energy source appears to be low. However, PUFAs account for almost half of this lipid fraction, with much of it occurring in the form of omega-3 (n-3) and omega-6 (n-6) fatty acids such as eicosapentanoic acid (EPA) and arachidonic acid (AA). PUFAs regulate a wide range of functions in the body including blood pressure, blood clotting, and correct development and functioning of the brain and nervous systems. Furthermore, PUFAs have a role in regulating inflammatory responses through the production of inflammatory mediators termed eicosanoids.

The n-3 to n-6 ratio of macroalgae is closely matched which may add to their efficacy as a dietary supplement or as part of a balanced diet. Moreover, they contain many essential fatty acids. Red and brown algae, for instance, are particularly rich in the n-3 fatty acids, EPA and α-linolenic acid, and the n-6 fatty acids, AA and linoleic acid, along with relatively high levels of oleic and palmitic acids. In contrast, green seaweeds, like.*Ulva pertusa*, are characterized by the presence of hexadecatetraenoic (n-3), oleic and palmitic acids. The n-3 fatty acid, octadecatetraenoic acid, is abundant in. *Laminaria*. sp. and.*Undaria pinnatifida* while hexadecatetraenoic acid is prominent in *Ulva* sp. In addition to fatty acids, the unsaponifiable fraction of macroalgae contains carotenoids (such as β-carotene, lutein and violaxanthin in red and green seaweeds, fucoxanthin in brown seaweeds), tocopherols, sterols (such as fucosterol in brown seaweeds) and terpenoids (Wall et al., 2010).

Polysaccharides

Although algal carbohydrate content is relatively high, macroalgae cannot be considered a potential energy rich food as digestibility of these carbohydrates is low. Moreover, the carbohydrate type varies greatly between algae species. Typical polysaccharides in red algae varieties consist of floridean starch, cellulose, xylan and mannan, and the water soluble fibre fraction is formed by sulfur containing galactans such as agar and carrageenan. Standard polysaccharides in brown algae are fucoidan, laminaran, cellulose, alginates and mannitol whilst the fibres are mainly cellulose and insoluble alginates. Most of these polysaccharides are not digestible by the human gastrointestinal tract and, therefore, can be regarded as dietary fibres. The total dietary fibre content of seaweeds ranges from 29.3–62.3 g/100 g and so is higher than the fibre content of most fruits and vegetables. Human consumption of algal fibre has been proven to be health promoting and its benefits are well documented.

Storage polysaccharides, such as agar, carrageenans and alginates, are the most commercially exploited components in seaweeds. These storage polysaccharides exhibit textural and stabilizing properties thus they are used in food applications such as thickening aqueous solutions, forming gels, forming water soluble films and stabilizing products such as ice-cream.

Fucoidans are a complex series of sulfated polysaccharides found widely in the cell walls of brown macroalgae. Fucoidans are reported to display numerous physiological and biological properties, including anticoagulant, antiviral, antithrombotic, antitumor and antioxidant activities, as well as having an effect on the inflammatory and immune systems. In addition, the therapeutic potential of fucoidans increases with their degree of sulfation and they can be easily extracted using either hot water or an acid solution. Another sulfated polysaccharide, porphyran, makes up the main components of the red macroalga,.*Porphyra*. This polysaccharide has reported uses as a gelling agent, a nutritional supplement and as an antioxidant. Alternatively, laminarin, the second major storage glucan in brown algae, has been identified as a modulator of intestinal metabolism through its effects on mucus composition, intestinal pH and short chain fatty acid production.

Another group of carbohydrate is oligosaccharides can be produced naturally or may be derived from algal polysaccharides after chemical, physical or biochemical degradations. Oligosaccharides can be beneficial to health when they are added to the diet to enhance the growth of prebiotic bacteria.

In this case, oligomers that resist the digestive process are used as a specific substrate for the growth of health beneficial bacteria. For instance, xylo-oligosaccharides and fructo-oligosaccharides are non-digestible oligomers that cannot be absorbed in the gastrointestinal tract. Hence, they are intact in the large bowel and are used as a preferential substrate by anaerobic bacteria such as bifidobacteria and lactobacilli. Interestingly, no specific conformation is correlated to the non-digestible oligosaccharide's biological activity, whereas the immunostimulating, antioxidant, antiangiogenic and antithrombotic activities of poly/oligosaccharides molecules are determined by glycan conformation (Courtois 2009).

Vitamins, Minerals and Antioxidants

One of the principal nutritive characteristics of seaweeds is their high antioxidant content (Table 1). In addition, vitamin B_{12}.is found in red macroalgae (e.g., *Palmaria longat*.and.*Porphyra tenera*) and in certain green seaweeds. Red and brown algae contain high levels of folic acid and folate derivatives including 5-metil-tetrahydro-folate, 5-formyl-tetrahydro-folate and tetrahydro-folate. Indeed, its amounts as high as 150 μg folic acid per 100 g of dry.*Undaria pinnatifida*.algae have been detected. As well as seasonal, environmental and physiological variations, vitamin content also depends on the type of seaweed processing. For example, the content of α-tocopherol in.*Himanthalia longate*. dehydrated (33.3 μg/g dry weight) was considerably higher than in canned. *Himanthalia longate* (12 μg/g dry weight) (Sánchez-Machado et al., 2002).

Table 1: Algal sources of dietary antioxidants

Antioxidant	Algal species	Reported levels (μg/g dry wt)
Vitamin C	*Ulva* sp.	94.20–1250
	Monostroma undulatum	1590–4550
	Undaria pinnatifida	1847.38
Vitamin E	*Ulva rigida*	19.70
	Ascophyllum nodosum	3.63
	Dunaliella tertiolecta	200–500
α-tocopherol	*Porphyridium cruentum*	55.2
	Laminaria ochroleuca	8.9 ± 2.1
	Tetraselmis suecica	190–1080
γ-tocopherol	*Porphyridium cruentum*	51.3
Carotenoids	*Porphyridium cruentum*	1020 ± 140
α-carotene	*Chlorella pyrenoidosa*	4232.50
	Dunaliella salina	2410–2690

Antioxidant	Algal species	Reported levels (μg/g dry wt)
β-carotene	*Ascophyllum nodosum*	
	Chlorella pyrenoidosa	4314.3
	Chlorella vulgaris	80–500
	Chlorococcum	
Antheraxanthin	*Dunaliella salina*	
	Laminaria digitata	
	Laminaria saccharina	
Astaxanthin	*Chlorella vulgaris*	
	Chlorococcum.sp.	
	Haematococcus pluvialis	up to 3%
β-cryptoxanthin	*Chlorella pyrenoidosa*	334.9
Cantaxanthin	*Chlorella vulgaris*	
	Chlorococcum	
Echinenone	*Phormidium*.sp.	
	Synechocystis.sp.	240
Fucoxanthin	*Ascophyllum nodosum*	
	Fucus serratus	
	Fucus vesiculosus	
Loroxanthin	*Chlorella pyrenoidosa*	
Lutein	*Chlorella protothecoides*	4600
	Chlorella pyrenoidosa	1153009.70
	Chlorella vulgaris	2970–3830
myxoxanthophyll	*Synechocystis*.sp.	580
neoxanthin	*Ascophyllum nodosum*	
	Chlorella pyrenoidosa	199.7
	Dunaliella salina	
violaxanthin	*Ascophyllum nodosum*	
	Chlorella pyrenoidosa	38.1
	Fucus serratus	
	Himanthalia elongata	50
	Laminaria digitata	
	Laminaria saccharina	
	Pelvetia canaliculata	
	Phormidium.sp.	
zeaxanthin	*Ascophyllum nodosum*	
	Chlorella pyrenoidosa	2170.3
	Dunaliella salina	11270 ± 1580
	Synechocystis.sp.	1640
Chlorophylls	*Dunaliella salina*	26–3100
	Himanthalia elongata	

Antioxidant	Algal species	Reported levels (µg/g dry wt)
chlorophyll.*a*	*Chlorella pyrenoidosa*	
	Chlorella vulgaris	3320–9630
	Chlorococcum	
chlorophyll.*b*	*Chlorella pyrenoidosa*	
	Chlorella vulgaris	2580–5770
	Chlorococcum	
pheophytin.*a*	*Chlorella vulgaris*	
	Porphyridium cruentum	3310 ± 1110
pheophytin.*b*	*Chlorella vulgaris*	2310–5640
	Porphyridium cruentum	30 ± 90
Polyphenols	*Fucus*.sp.	41400 ± 400
	Haematococcus pluvialis	
	Laminaria.sp.	7300 ± 100

Microalgae

Microalgae are the most primitive and simply organized members of the plant kingdom, with the majority existing as small cells of about 3–20 µm. These algae are ubiquitous in nature and aquatic microalgae have been isolated in areas ranging from hot springs to glacial ice flows. Microalgae are found in both benthic and littoral habitats and also throughout the ocean waters as phytoplankton. Phytoplankton comprises organisms such as diatoms (bacillariophyta), dinoflagellates (dinophyta), green and yellow-brown flagellates (chlorophyta; prasinophyta; prymnesiophyta, cryptophyta, chrysophyta and rhaphidiophyta) and blue-green algae (cyanophyta). As photosynthetic organisms, this group plays a key role in the productivity of oceans and constitutes the basis of the marine food chain. There are over 50,000 different species of microalgae of which only a few have been characterized. This group of microorganisms is extremely diverse and represents a major untapped resource of valuable bioactive compounds and biochemicals such as pigments, antioxidants, polysaccharides, sterols, fatty acids and vitamins (Bhakuni and Rawat 2005).

Proteins, Peptides and Amino Acids

The high protein content of various microalgal species and their amino acid pattern, which compares favorably with that of other food proteins, is a good endorsement of microalgae as an alternative protein source. *Spirulina*, for instance, is high in protein (60–70% depending on the strain) and, not only does this protein possess all of the essential amino acids, but these amino acids have excellent bioavailability. Furthermore, the industrial scale growth of the microalga, *Dunaliella*, can turn out protein extract at about 100 times greater productivity than that reported in agriculture and 50 fold greater than in fish farming. Proteins from marine sources show promise as functional ingredients in foods because they possess numerous important and unique properties such as film and foaming capacity, gel forming ability and antimicrobial activity. In addition, purified peptides from *Chlorella vulgaris*.have demonstrated significant protective effects against cellular damage. With regard to one of the major proteins in *Spirulina platensis* and *Porphyridium*, phycobiliprotein, several therapeutic bioactivities have been described, namely, hepatoprotective, anti-inflammatory, immunomodulating, antioxidant and anticancer effects (Spolaore et al., 2006).

Fatty Acids

The average lipid content of algal cells varies between 1 and 70% but can reach 90% of dry weight under certain conditions. Algal lipids are composed of glycerol, sugars or bases esterified to saturated or unsaturated fatty acids. Among all the fatty acids in microalgae, some fatty acids of the n-3 and n-6 families are of particular interest. According to Mendes et al (1995) the main constituents of the lipidic fractions of.*Chlorella vulgaris*.are oleic, palmitic and linolenic acids, accounting for 41, 22 and 9% of the total amount, respectively. Additionally, palmitic, linolenic and oleic acids account for more than 85% of the total fatty acid content of.*Dunaliella salina*.while the green microalga,.*Haematococcus,* has been shown to contain short chain fatty acids with antimicrobial activity (Mendes et al., 1995).

Higher plants and animals lack the requisite enzymes to synthesize PUFAs of more than 18 carbons and so have to obtain them from their food. Fish and fish oil are the common sources of long chain PUFAs but safety issues have been raised because of the possible accumulation of toxins in fish. Moreover, the application of fish oil as a food additive is limited due to problems associated with its typical fishy smell, unpleasant taste and poor oxidative stability. Consequently, long chain PUFAs are commercially produced via microalgae cultivation for incorporation into infant milk formulations and for use as dietary supplements and food additives. Maximum n-3 fatty acid production can also be induced by altering the growth conditions of microalgae. For instance, under optimal culture conditions, *Chlorella minutissima* can produce an EPA content of up to 45% of its total fatty acid content.

Microalgae such as *Porphyridium,* which shows a relatively low lipid content, contains significant amounts of several major fatty acids such as palmitic acid, AA, EPA and linoleic acid *Spirulina* provides an interesting source of γ-linolenic acid (20–25% of the total lipid fraction), which is a precursor of prostaglandins, leukotrienes and thromboxans involved in the modulation of immunological, inflammatory and cardiovascular responses. This microalga is also a natural source of active fatty acids such as lauric, palmitic and oleic acids with the n-3 fatty acid, docosahexaenoic acid (DHA), accounting for up to 9.1% of the total fatty acids content (Spolaore et al 2006).

Polysaccharides

Carbohydrates in microalgae can be found in the form of starch, glucose, sugars and other polysaccharides. Their overall digestibility is high, which is why there is no limitation to using dried whole microalgae in foods or feeds. Moreover, the biological activities of some microalgal species have been associated with polysaccharides. Polysaccharide complexes from. *Chlorella pyrenoidosa*, and possibly. *Chlorella ellipsoidea*, contain glucose and any combination of galactose, rhamnose, mannose, arabinose,.*N*-acetylglucosamide and *N*-acetylgalactosamine. These complexes are believed to have immunostimulating properties, specifically immune stimulatory activity and can inhibit the proliferation of.*Listeria monocytogenes*.and. *Candida albicans*. The most important substance in *Chlorella*.is β-1, 3-glucan, which is an active immunostimulator, a free radical scavenger and a reducer of blood lipids. *Chlorella* can also be used as a food additive owing to the taste and flavour adjusting actions of its coloring agent. Also, novel polysaccharides isolated from *Porphyridium* and *Nostac flegelliforme*.microalgae exhibited potent antiviral activity against herpes simplex virus (HSV-1 and 2) both *in vitro* and *in vivo* (Huheihel et al., 2002).

Antioxidants

The nutritional and therapeutic relevance of dietary carotenoids is attributed to their ability to act as provitamin A; that is, they can be converted into vitamin A by the human body. Moreover, carotenoids play a protective role by preventing the formation of reactive oxygen species. Microalgal production of carotenoids, such as β-carotene and astaxanthin, is an attractive area of research as they are valuable bioactive ingredients that can present at relatively high concentrations in algal cells (Table 1). Moreover, cultivated algae can be induced to produce even larger quantities of carotenoids by controlling certain environmental growth conditions. The strains of microalgae that are currently being investigated for use as natural producers of commercial carotenoids include. *Dunaliella salina, Sarcina maxima, Chlorella protothecoides, Chlorella vulgaris* and *Haematococcus pluvialis*.

Dunaliella salina is the most suitable organism for the mass production of β-carotene as it can produce β-carotene up to 14% of its dry weight. This microalga can also be cultivated easily and quickly when compared to plants and produces both *cis* and *trans* isomers of carotenoids for high bioavailability and bioefficacy. B-carotene is one of the leading food colorants in the world and has been applied to a range of food and beverage products to improve their appearance to consumers. In addition, β-carotene has strong antioxidant properties which help to mediate the harmful effects of free radicals implicated in numerous life-threatening diseases, including various forms of cancer, coronary heart disease (CHD), premature aging and arthritis. The antioxidant qualities of β-carotene can also assist the body in suppressing the effects of premature aging caused by UV rays (Miyashita 2009).

Haematococcus.is another unicellular alga that can be used in both open and closed culture systems for the production of antioxidants, namely chlorophylls and carotenoids. Under stress conditions, *Haematococcus pluvialis* has the ability to accumulate large quantities (1.5–3% of dry weight) of the high value carotenoid, astaxanthin. Besides, the United States Food and Drug Administration for marketing have cleared.*Haematococcus pluvialis*.as a dietary supplement and it has also been approved in several European countries for human consumption. With an antioxidant activity up to 10 times stronger than other carotenoids, astaxanthin provides protective activity against cancer, inflammation and UV light. The health benefits of astaxanthin along with its strong colouring properties make it a potential ingredient for use in the nutraceutical, cosmetics, food and feed industries. It has clearly been established that microalgae are a rich source of nutritious and biologically active compounds, namely carotenoids, phycobilins, fatty acids, polysaccharides, vitamins and sterols. Nevertheless, not only is it their huge diversity that makes these microorganisms interesting, but also the possibility of growing them at different conditions and using them as natural reactors, leading to an enrichment of some bioactive compounds. However, prior to this, algal material must be analyzed for the presence of toxic compounds (Guerin et al., 2003).

Potential to Reduce Prevalence of Chronic Diseases

The increasing number of scientific papers published in the last two decades correlating diet and some chronic diseases have shown the extraordinary possibilities of foods to support, or even to improve, our health. As a consequence, there is now a huge interest among consumers and the food industry on products that can promote health and well-being. The marine world represents a largely untapped reserve of bioactive ingredients and considerable potential exists for exploitation of these compounds as functional food ingredients (Table 2). Substances such as chitin, chitosan, n-3 oils, algae, carotenoids, vitamins and minerals, calcium in fish bone, bioactive peptides and fish protein hydrolysates provide a myriad of health benefits, including reduction of CHD, anticarinogenic and anti-inflammatory activities.

Table 2: Potential marine functional food ingredients and their health benefits

Functional food ingredient	Health benefit	Marine source
Peptides	ACE inhibition	Fish frame, algae
	Anticoagulative	Fish frame
	Antidiabetic	Fish frame
	Antimicrobial	Marine invertebrates, fish
	Antioxidative	Algae protein waste, fish frame
n-3 fatty acids	Anticarcinogenic	Fish
	Anti-inflammatory	Fish, mussels
	Cardioprotective	Fish
	Cognitive function	Fish
Polysaccharides	Anticarcinogenic	Algae, crustaceans (chito-oligosaccharides)
	Antioxidative	Algae, crustaceans (chito-oligosaccharides)
	Antiviral	Algae
	Cardioprotective	Algae
Carotenoids	Anticarcinogenic	Algae
	Antioxidative	Algae
	Anti-obesity	Algae
	Antidiabetic	Algae
Chlorophyll	Anticarcinogenic	Algae
Polyphenols	Antidiabetic	Algae
	Antioxidative	Brown algae

Cancer

Many potent natural products which display effective anticancer activities have been discovered in the marine environment. Indeed, since the early 1990s, there has been a dramatic increase in the number of preclinical anticancer lead compounds from marine sources that have entered into human clinical trials. One of these compounds, trabectidin (Yondelis®), originally isolated from the Caribbean marine tunicate.*Ecteinascidia*.turbinate, has been approved for use as an anticancer agent in Europe. The lichen metabolites, sphaerophorin and pannarin, prevented UV light and nitric oxide mediated plasmid DNA damage, and attenuated the growth of melanoma cells by, at least in part, triggering an apoptotic process. Moreover, a number of isolated marine sponge compounds are inhibitors of protein kinase C (PKC) e.g., BRS1, isoaaptamine, debromohymenialdisine. PKC inhibitors have attracted much attention as there is evidence that too high levels of PKC enzyme are involved in the pathogenesis of arthritis and psoriasis, and in tumor development. PKC is believed to be the receptor protein of tumor promoting phorbol esters, and PKC inhibitors prevent binding of carcinosarcoma cells to the endothelium.The cytoskeleton is also an interesting target for cancer therapy, as the microtubules and microfilaments are involved in cellular organisation during cell division. A number of extracts of marine sponges and ascidians are believed to inhibit the protein by binding to the microtubule binding site, "locking up" the protein's motor function, and thereby blocking cell division (Mayer and Gustafson 2008).

Algal Polysaccharides

Non-digestible oligosaccharides are known to improve gut microecology and in doing so may reduce the risk of colon cancer. Intake of transgalactosylated disaccharides reduces the faecal pH as well as ammonia, *p*-cresol and indole concentrations, with an increase in *Bifidobacteria* and Lactobacilli and a decrease in *Bacteroidaceae*.populations. As these changes in faecal physiological parameters are believed to reduce the risk of cancer development, macroalgal non-digestible oligosaccharides could be considered as potential anticarcinogenic food ingredients (Mussatto and Mancilha 2007).

Carotenoids and Chlorophylls

Being lipid soluble, carotenoids are absorbed with fats and circulate bound to different lipoproteins. The principal biological effects of carotenoids relate to their antioxidant properties, which form the basis of potential protection against lipid peroxidation, atherogenesis, DNA oxidation, and cancer. Carotenoids have been implicated in the inhibition of cancer cells.*in vitro*. in animal models and in humans, as important dietary phytonutrients having cancer preventive activity for lung, colon, breast and prostate cancer. The food sources that yield chlorophyll derivatives may play a significant role in cancer prevention. The dietary chlorophyll derivatives exhibit antimutagenic effects and reduce tumor cell growth. Indeed, epidemiological evidence has linked diets high in chlorophyll with a reduced risk of colon cancer in humans. Antioxidant activity, mutagen trapping, modulation of detoxification pathways, and induction of apoptosis in cancer cells have been highlighted as possible modes of actions responsible for chlorophyll's protective effects.*in vivo*.(Ferruzzi and Blakeslee 2007).

Cardiovascular Disease

Cardiovascular disease (CVD) is a class of diseases that affect the heart, blood vessels (arteries and veins) and blood circulation, and is one of the leading causes of mortality and morbidity worldwide. Examples of CVD include atherosclerosis, CHD, stroke, heart failure, deep vein thrombosis and peripheral arterial disease. In relation to marine bioactives, there is considerable evidence that these compounds can help to reduce the risk factors associated with CVD. Low density lipoprotein (LDL) cholesterol was significantly lower in rats fed a diet containing dried.*Ulva rigida*.while Oben and colleagues found that individuals given a freshwater algae infusion displayed lower total cholesterol, LDL cholesterol and triglyceride levels, and higher high density lipoprotein (HDL) cholesterol values than those given a water placebo (Oben et al., 2007).

Arthritis

Arthritis describes a condition involving inflammation of the joints and is a disease affecting mostly the aged population. Preventing inflammation with its associated pain and reduced mobility symptoms is a primary requirement in arthritis treatment. Due to the involvement of inflammatory eicosanoids in the aetiology of this disorder, diet may be a potential therapeutic agent and so the importance of dietary PUFAs in the treatment of arthritis has been greatly investigated. Meta-analysis of 17 randomised, controlled trials was conducted by Goldberg and Katz to assess the analgesic effects of n-3 PUFAs in patients with rheumatoid arthritis or joint pain. Favourable outcomes such as reduced patient reported joint pain intensity, minutes of morning stiffness; number of painful and/or tender joints and reduced non-steroidal anti-inflammatory drugs consumption were reported following 3 to 4 months supplementation with n-3 PUFAs. In fact, the authors conclude that n-3 PUFA supplementation is an attractive adjunctive treatment for joint pain (Goldberg and Katz 2007).

Asthma

Fish oil or fish containing more than 2% fat has been found to have a reduced risk of airway hyper responsiveness, and children who regularly eat fresh, oily fresh have a reduced risk of developing asthma than children who rarely eat fish. Supplementation of diet with n-3 fatty acids also confirmed their benefit in the reduction of breathing difficulties and other symptoms, along with reduced drug doses required by asthma patients. During treatment, the increase in the content of n-3 fatty acids in cell membranes was reported to take place at the expense of AA resulting in the competitive inhibition of pro-inflammatory eicosanoid production and production of anti-inflammatory eicosanoids (Hodge et al., 1996).

Neuroinflammation

Recently, there has been recognition of an inflammatory component to the pathology of neurodegeneration, most notably in Alzheimer's disease but also in Parkinson's disease and motor neuron disease. As anti-inflammatory n-3 PUFAs are preferentially incorporated in the brain, a diet rich in EPA and

DHA could keep neuroinflammation at a minimum. In fact, elderly people who eat fish or seafood, that are highly enriched in n-3 PUFAs, at least once a week, have been shown to be at lower risk of developing dementia including Alzheimer's disease. Evidence is also emerging which suggests that marine algae could possess therapeutic activities for combating neurodegenerative diseases associated with neuroinflammation. Red alga *Neorhodomela aculeate* could be considered as a potential neuroprotective and anti-inflammatory agent to treat aging related neurological disorders (Lim et al., 2006).

Cognitive Decline and Depression

Cognitive impairment and dementia are frequent disorders among elderly persons and influence the individual's ability to function independently. Due to the aging of the population, the prevalence of cognitive impairment and dementia are expected to increase. Drugs currently used in the treatment of cognitive decline and dementia have a very limited therapeutic value, suggesting the necessity to potentially individualize new strategies able to prevent and to slow down the progression of predementia and dementia syndromes. Numerous epidemiological studies on the association between diet and cognitive decline suggest a role of fatty acids intake in maintaining adequate cognitive functioning and possibly in preventing or delaying the onset of dementia, both of degenerative or vascular. In particular, fatty fish and marine n-3 PUFA consumption was associated with a reduced risk. Moreover, a diet enriched with the algae,.*Chlorella*, reduced oxidative stress and significantly prevented the decline of cognitive ability in an age dependent dementia mouse model. The authors suggest that the prolonged consumption of.*Chlorella*.has the potential to prevent the progression of cognitive impairment (Nakashima Y, et al., 2009).

In general, PUFAs of marine origin appear to be suitable candidates for functional food ingredients to relieve memory deficits associated with aging. Promising results have already been reported in young population samples. Supplementation with a fish-flour bread spread containing n-3 PUFAs, embedded in a natural food matrix, had a beneficial effect on learning and memory of children. In addition, a study involving marine collagen peptide (MCP) derived from Chum Salmon skin demonstrated that MCP facilitates learning and memory in aged mice by reducing oxidative damage in the brain and increasing brain derived neurotrophic factor and PSD95 protein expression (Pei et al., 2010).

Diabetes

Dietary management of diabetes involves maintaining both blood glucose and blood lipid concentrations at as near normal levels as possible to reduce the possibility of associated complications developing, e.g., CVD, diabetic neuropathy, retinopathy and nephropathy. As well as offering protection against heart disease and cancer, some marine extracts have been associated with improvements in glycaemic control.

Table 3: Anti-diabetic properties of marine natural products with established mechanisms of action

Compound	Source	Experimental model	Effect/Mechanism of action
α-galactosylceramide	*Agelas mauritianus*. sponge	Non-obese diabetic mice	Suppression of IFN-γ, increase of serum Ig E levels, and promotion of islet autoantigen specific Th2 cells Suppression of T- and B-cell autoimmunity to islet beta cells
Aqueous extracts	*Xetospongia muta*. sponge,.*Bunodosoma granulifera*and. *Bartholomea annulata*sea anemones	*In vitro* models	Inhibition of dipeptidyl peptidase IV activity
Ethanolic extract	*Ulva rigida*.alga	Wistar diabetic rats	Decreased blood glucose concentrations
Extract	*Posidonia oceanica*phanerogam	Wistar diabetic rats	Decreased blood glucose concentrations
Fucosterol	*Pelvetia siliquosa*.alga	Sprague-Dawley diabetic rats	Reduction in serum glucose concentration and inhibition of sorbitol accumulation in the lenses
Marine collagen peptides	Wild fish	Human diabetic subjects	Decreased free fatty acids, cytochrome P450 and hs-CRP Regulation on metabolic nuclear receptors
Methanolic extract	*Ecklonia cava*.alga	Sprague-Dawley diabetic rats	Reduction in plasma glucose levels and increased insulin concentration Activation of AMPK/ACC and PI3/Akt signalling pathways

Compound	Source	Experimental model	Effect/Mechanism of action
Microalgal extracts	*Chlorella*.sp. alga,. *Nitzschia laevis*. diatom	*In vitro* models	Inhibition of advanced glycation endproducts (AGEs) formation
n-3 PUFAs	Fish oil	Wistar rats	Restoration of insulin receptor and insulin receptor substrate-1 tyrosine phosphorylation Maintenance of phosphatidylinositol-3' kinase activity and GLUT-4 content in muscle
	Fish oil	Healthy human subjects	Reduction in glucose oxidation, increased fat oxidation and glycogen storage
Phenylmethylene hydantoins	*Hemimycale arabica*sponge	*In vitro*.model Sprague-Dawley rats	Inhibition of glycogen synthase kinase-3β activity Increased liver glycogen
Phlorotannin components	*Ascophyllum nodosum*.alga	*In vitro* models	Inhibition of α-amylase and α-glucosidase activities
Sodium alginate	*Laminaria angustata*. alga	Wistar rats	Inhibition of rising blood glucose and insulin levels

Conclusion

With increased life expectancy, our diet will play a key role in sustaining human health. This is a challenge for the food industry as consumers not only demand tasty and convenient food, but also healthy, nutritious food. Marine nutraceuticals are both a coherent and attractive option for the food industry as there are a multitude of functional food ingredients that can be derived from marine sources. It is also believed that, as people become increasingly aware of the association between diet and good health, the consumption of fishery products will most likely increase. Moreover, in relation to marine algae, their content in proteins, carbohydrates, lipids, fibre, metabolites,.*etc.*can be influenced by their growing parameters (water temperature, salinity, light and nutrients). This means that, from a biotechnological perspective, algae can be considered as natural bioreactors, able to provide different types of compounds at different quantities—an appealing attribute to the functional food industry.

Despite the vast possibilities for use of marine bioactives in food, more multidisciplinary research is needed. All aspects, including chemical composition, biotechnology, extraction, bioactivity, and toxicity should be considered. Also, to effectively transfer the research into a practical field in which marine bioactives can be used as functional food ingredients, the exact functional activity of the extracts/compounds should be determined. This primary objective should be based on investigations *in vitro* or *ex vivo* in cellular lines or culture tissues, later in animal models and then corroborated in studies of observation or intervention in human clinical trials. Finally, to be employed as ingredients in food products, different studies should be carried out to determine if their activity is maintained after manufacturing and cooking processes.

In conclusion, marine bioactives appear to fit the criteria for development as functional food ingredients. Firstly, they are widely available, with a guaranteed supply. Secondly, marine bioactives are naturally occurring compounds, and their isolation/extraction is relatively cost effective. Lastly, and probably most importantly, they are functional—their biological activities affect the pathogenesis of several diseases. Consequently, ongoing efforts should be made into the research and development of marine functional foods with prospect that, in the future, their consumption could lead to a reduction in the prevalence and severity of chronic diseases.

References

Aneiros A, Garateix A. Bioactive peptides from marine sources: Pharmacological properties and isolation procedures.J Chromatogr B.2004; 803:41–53.

Bhakuni DS, Rawat DS. Bioactive Marine Natural Products.1st ed. Anamaya Publishers; New Delhi, India: 2005.

Bocanegra A, Bastida S, Benedí J, Ródenas S, Sánchez-Muniz FJ. Characteristics and nutritional and cardiovascular-health properties of seaweeds.J Med Food.2009; 12:236–258.

Courtois J. Oligosaccharides from land plants and algae: Production and applications in therapeutics and biotechnology.Curr Opin Microbiol.2009; 12: 261–273.

Ferruzzi MG, Blakeslee J. Digestion, absorption, and cancer preventative activity of dietary chlorophyll derivatives.Nutr Res.2007; 27:1–12.

Fleurence J. Seaweed proteins: Biochemical, nutritional aspects and potential uses.Trends Food Sci Technol.1999; 10:25–28.

Goldberg RJ, Katz J. A meta-analysis of the analgesic effects of omega-3 polyunsaturated fatty acid supplementation for inflammatory joint pain.Pain.2007;129:210–223.

Guerin M, Huntley ME, Olaizola M. *Haematococcus astaxanthin*: Applications for human health and nutrition.Trends Biotechnol.2003;21:210–216.

Hodge L, Salome CM, Peat JK, Haby MM, Xuan W, Woolcock AJ. Consumption of oily fish and childhood asthma risk.Med J Aust.1996; 164:137–140.

Honkanen P. Consumer acceptance of (marine) functional food. In: Luten J, editor.Marine Functional Food.1st ed. Vol. 1. Wageningen Academic Publishers; Wageningen, The Netherlands: 2009. pp. 141–154.

Huheihel M, Ishanu V, Tal J, Arad S. Activity of*porphyridium*.sp. Polysaccharide against herpes simplex viruses.*in vitro*.and.*in vivo*.J Biochem Biophys Methods.2002; 50: 189–200.

Lim CS, Jin D-Q, Sung J-Y, Lee JH, Choi HG, Ha I, Han J-S. Antioxidant and anti-inflammatory activities of the methanolic extract of *Neorhodomela aculeate*. in hippocampal and microglial cells.Biol Pharm Bull.2006; 29: 1212–1216.

Mayer AMS, Gustafson KR. Marine pharmacology in 2005–2006: Antitumour and cytotoxic compounds. Eur J Cancer. 2008; 44: 2357–2387.

Mendes RL, Fernandes HL, Coelho JP, Reis EC, Cabral JM, Novais JM, Palavra AF. Supercritical CO_2 extraction of carotenoids and other lipids from *Chlorella vulgaris*. Food Chem. 1995; 53: 99–103.

Miyashita K. Function of marine carotenoids.Forum Nutr.2009; 61: 136–146.

Mussatto SI, Mancilha IM. Non-digestible oligosaccharides: A review. Carbohydr Polym. 2007; 68: 587–597.

Nakashima Y, Ohsawa I, Konishi F, Hasegawa T, Kumamoto S, Suzuki Y, Ohta S. Preventive effects of chlorella on cognitive decline in age-dependent dementia model mice. Neurosci Lett.2009; 464: 193–198.

Oben J, Enonchong E, Kuate D, Mbanya D, Thomas T, Hildreth D, Ingolia T, Tempesta M. The effects of proalgazyme novel algae infusion on metabolic syndrome and markers of cardiovascular health.Lipids Health Dis.2007; 6: 20.

Pei X, Yang R, Zhang Z, Gao L, Wang J, Xu Y, Zhao M, Han X, Liu Z, Li Y. Marine collagen peptide isolated from chum salmon (oncorhynchus keta) skin facilitates learning and memory in aged c57bl/6j mice.Food Chem.2010; 118:333–340.

Rasmussen RS, Morrissey MT. Marine biotechnology for production of food ingredients.Adv Food Nutr Res.2007;52: 237–292.

Sánchez-Machado D, López-Hernández J, Paseiro-Losada P. High-performance liquid chromatographic determination of [alpha]-tocopherol in macroalgae.J Chromatogr A.2002; 976: 277–284.

Spolaore P, Joannis-Cassan C, Duran E, Isambert A. Commercial applications of microalgae.J Biosci Bioeng.2006; 101: 87–96.

Wall R, Ross RP, Fitzgerald GF, Stanton C. Fatty acids from fish: The anti-inflammatory potential of long-chain omega-3 fatty acids.Nutr Rev.2010; 68:280–289.

12

Functional Food Ingredients in Milk and Milk Products

[1]Gaurav Kr Deshwal, [2]Swati Tiwari, [1]Heena Sharma, [1]Narender Raju Panjagari, [1]Ashish Kumar Singh, [3]Priti Khemariya

[1]ICAR-National Dairy Research Institute, Karnal, Haryana
[2]Department of Food Science and Technology, Pondicherry University Puducherry
[3]Indian Institute of Packaging, Delhi

Introduction

In recent years, food scientists, nutritionists and health experts have become more inclined towards functional food components present in various food sources due to their positive influence on health, which is beyond providing the basic nutrition.Nowa days, generalconsumers are also seeking healthier food options because of nutritive food habits and awareness. Therefore, it has triggered the studyof food and deep mining of food components from nutritional and health benefit's perspective also. Functional food ingredient can be defined as any componentwhich is more than a source of nutrient i.e. provides additional specific health benefits. Milk is well known to exert positive effect on human health, provides nutrition and widely consumed all across the globe in form of either milk or milk products. As per the definition, milk is heterogeneous blend of lacteal secretion which contains various elements that displays wide range of chemical and functional activities. It has been more than a source nutrition and supplement to different age groups either neonates, youngsters or grown up peoples (Park, 2009). It is one of the basicelements of many diets either in its native form or fermented form which every class of people can afford and had evolved as one of the important nourishment sources. Milk in its native form contains fewer bioactive components while fermentation process produces numerous secondary metabolites which possess various health-beneficial properties (Stanton et al., 2005).

Dairy sector holds the major proportion of the overall functional food market. Specifically, functional dairy foods such as probiotics and prebiotic based dairy drinks, yoghurts, chesses and other cultured dairy products have central role in growing functional food's market. This has been proven inpast few decades that milk is rich sourceof various functional and bioactive components thereby creating its application as nutraceutical, functional food along with various medicinal applications. The most important functionalconstituents of milk are proteins orspecifically biologically active peptides. These bioactive peptides possess various health beneficial effects including immunomodulatory, antihypertensive and can even provide antibacterial assurance against enteropathogens. These expanded utility and usefulness of milk can be found even in protein constituents alone.Proteins comprise a variety of serum and glandular compounds which are distinctive in atomic size, concentration and functionality (Regester et al., 1997). Not only proteins, other constituents of milk such as milk carbohydrates, vitamins, minerals and certain other growth factors are also equally important from the view point of functional foods. Milk oligosaccharides such as galacto-oligosaccharides (GOS) and fructo-oligosaccharides (FOS) enhance the intestinal microbiota, subsequently immune response which in turn plays a key role against several allergies and infections. Apart from it, lipid components of milk are also very important in this perspective and several efforts have been made to enrich the milk with bioactive lipid components including different fatty acids (Mills et al., 2011) such as MUFA (Monounsaturated Fatty Acids), SFA (Short Chain Fatty Acids) and CLAs (conjugated Linoleic Acid).A summarised version of different functional milk ingredients and their health benefits or therapeutic roles are presented in Table 1 while Figure 1 represents the effect of different milk bioactive components on human body.

Table 1: Therapeutic factors, their roles and milk components responsible for the bioactive activities

Key factors	Key functions	Responsible milk components
Antihypertensive	Reduces stress, reduction of blood pressure	Casein-derived dodecapeptide, α_{s1}-casein
ACE inhibitory	Prevents angiotensin-II formation, a potent vasoconstrictor	β- and κ-casein.
Antioxidant	Maintains glutathione(GSH) activity which is a potent antioxidant system, one of the principal cellular protection mechanisms	Casein phosphopeptides, β-lactoglobulins

Key factors	Key functions	Responsible milk components
Immunomodulatory	Increased natural killer cell activity, thymocyte trafficking immunological development, and interleukins-6; decrease the activity of tumor necrosis factor-α	Lactoferrin, Immunoglobulins, Cytokines
Opioid Antagonist	Decreased gut mobility, gastric emptying rate; increased amino acids and electrolyte uptake	α_{s1}-, α_{s2},- β- and κ casein, α-lactalbumin
Antibacterial	Disturbs normal permeability of bacterial cells	Lactoferrin, Lactoferricins, Lactoperoxidase, Lysozyme
Mineral carrying	Carriers for different minerals by forming soluble organophosphate salts	Caseinophosphopeptides, Enzymatically hydrolyzed α_{s1}-, α_{s2}-, β- and κ-caseins.
Antithrombotic	Prevents thrombus formation in arteries, veins or the chambers in the heart	κ-casein
Cytomodulatory	It can modulate cell viability such as proliferation and apoptosis thereby can inhibit cancer cell growth	Caseinophosphopeptides.
Antiappetizing	Suppresses hunger thereby prevents obesity	Whey protein derived peptides
Bifidogenic	Enhances the growth of *Bifidobacteria*	Fructooligosachharides and Galactooligosachharides
Probiotic	Improved gut health	Oligo and Polysachharides

Milk Constituents Responsible for Biological Activity

Milk Proteins and Peptides: Paramount Functional Constituents of Milk

The protein present in milk contains numerous biologically active components. Mostly these substances are present in inactive form but they can be released from precursor proteins by their proteolysis. Several bioactive peptides from milk proteins (whey proteins, casein, lactoferrin, peptides and milk fat globule membrane proteins) are potential modulators of different regulatory processes in the body and subsequently brings positive influence on human health. Various milk proteins and peptides with functional properties are discussed hereafter:

Casein Derived Peptides

Caseinophosphopeptides (CPPs), mainly from α and β-caseins, are the predominant bioactive peptides with multiple functionalitiessuch as mineral binding and cytomodulatory properties. Organic phosphorus isvital for CPPs to work and the content of organic phosphorus is higher in human milk therefore the concept of 'humanising' bovine milk is in recent trend for the preparation of infant formulas. CPPs are of prime importance as they can form soluble organophosphate salts and may work as transporters for various minerals, particularly calcium, besides phosphorus, iron, zinc and copper. CPPsare the most important peptides in this category and had been proved to enhance calcium absorption by limiting the precipitation of calcium in the distal ileum (Meisel and Fitz Gerald, 2003) of rachitic infants. Mellander, 1950 also reported that calcium.bound to phosphopeptides promotes calcification without further administration of.vitamin D. In past decades, various studies have been conducted on animals to assess the effect of calcium absorptivity due to CPPs but limited numberof human studies have been conducted. The calcium absorption gets increased remarkably with CPPs in post-manopausal women during its co-ingestion (Heaney et al., 2994).

The second important activity exerted by CPPs is their cytomodulatory property. Casein related peptides have great potential as apoptosis inducing peptides for cancer cells, especially purified peptides are claimed to have effects on cell modulation, and therebypreventing proliferation and finally leads to apoptosis of cancer cells. The primary target site for apoptotic activity of cancer cells and control of malignant cell proliferation is gastrointestinal tract. This antiproliferative effects for cancer cells of milk-proteins derived peptides in colon cancer cell lines indicates that they can prevent development of cancer cells by promoting the apoptosis (Ganjam et al., 1997), but there is need to conduct more human trials for claiming its effect. Despite the fact of few conducted human trials, both mineral binding and cytomodulatory peptides related CPPs of milk are professed to be health enhancing components that can be utilized to lessen the risk of disease or to enhance certain physiological function.

κ-Casein

κ-Casein peptides doesn't show any significant role in antimicrobial or immunomodulatory activity. However, f103-111 fragment or peptide sequence of κ-caseinhad been reported to prevent blood clotting and κ-caseinoglycopeptides also serves as a source of antithrombotic peptides (Fiat et al., 1993). The f(108-110) peptide of κ-caseinanalogous to IPP (Ile-Pro-Pro), purified from a Japanese skim milk based fermented drink demonstrated ACE (Angiotensin Converting Enzyme) inhibitory activity or antihypertensive activity (Nakamura et al., 1995). Agonistic peptides of κ-casein f(35-41), f(58-61), f(25-34) and f(158-164), enhance the activity of casoxins (Xu, 1998).

κ-casein has been related with stimulatory effect on proliferation of human hybridoma HB4C5 cells and immunoglobulin production. The digestion of κ-casein with chymosin resulted in fractionation of para-κ-casein and GMP, both fragments stimulated proliferation of cells, but only para-κ-casein stimulated IgM (Immunoglobulin M) production (Kapila et al., 2008). Recently, peptide named κ-casecidin (17-21, Phe-Phe-Ser-Asp-Lys) was isolated from trypsin digests of bovine κ-casein and it showed cytotoxic activity towards mouse spleen cells and some mammalian cells including human leukemic cells.

Whey Proteins Derived Peptides

β- Lactoglobulins

It is classified as one of the predominant portion of whey protein which contributes upto 58% of the total whey proteins. The prime functional role provided by β-Lactoglobulin (β-Lg) is their immunomodulatory effects. Several in vitro and in vivo studies have indicated that such whey proteins can modulate the antibody production. These were known to enhance the production of IgM and also known to play important role in lymphocyte proliferation (Gauthier et al., 2006). Theyare also considered as opioid antagonist and have some radical scavenging activity hence they can be regarded as possible anti-oxidative peptides. A peptide sequence from β-Lg, tryptichydrolysate, was found to have a hypocholestrolemic effect which suppresses the cholesterol absorption in intestine (Nagoaka et al., 2001). The mechanism of hypocholestrolemic activity was mainly due to β-Lg, and their cholesterol lowering property was found to be greater than the medicine, β-sitosterol, in animal studies. β-lg bears two distinct ligand binding sites per molecule which can bind with hydrophobic ligands and in one of the studyreported its ability to bind with palmitic acid (Wu et al., 1999), highlighting its role in palmitic acid metabolism. Furthermore, it may often bind to small hydrophobic ligands such as retinol because it possess the structure similar to retinol binding protein (RBP), calcium and certain fatty acids (Cho et al., 1994).

α- Lactalbumin

α- Lactalbumin (α-LA) are one of the whey derived proteins that can reduce the risk of incidence of some cancers. This protein possesses antiproliferative effects in colon adenocarcinoma cell lines (Sternhagen & Allen, 2001). It was shown that they promote the apoptosis in cancer cells corresponding to transformed or immature cells with sparing nature on mature epithelial cells at the same time (Hakansson et al., 1995). They also exhibit opioid antagonist activity; whose characteristic structural feature is presence of tyrosine residue at terminal end with aromatic amino acids residues. As mentioned already, fermentation releases these peptides from their native form and releases opioid antagonist peptides in the form of α-LA f50-53 (Rokka et al., 1997). It was found that a diet rich in α-LA have role in increasing the ratio of plasma tryptophan to large neutral amino acids along with improved cognitive power (Markus et al., 2002).

Immunoglobulins

Immune system modulation is a chief role played by immunoglobulins (IGs) which is often regarded as anti-inflammatory antibody, especially in passive immunity. IGshave been demonstrated to serve protection against illnesses in newly born babies through passive immunity. It was found to have significant role for providing protection against pathogens at mucosal surfaces by serving as key element for fixing and maintaining homeostasis of intestine by innate and adaptive immunity mechanisms (Cakebread et al., 2015). Although, the amount of IGs in bovine milk is less as compare to human milk and it promotes the concept of humanising the bovine milk. Nevertheless, bovine derive antibodies are important for providing protection against dental caries caused by streptococci and imparts immunity against plaque formation (Loimaranta et al., 1999). IGs are also believed to have antimicrobial property to some extent.

Lactoferrin

Lactoferrin (Lf), which is an iron-binding.glycoprotein belongs to transferrin family, oftenly known as the red fraction of milk, has been recently termed as nutraceutical protein (Iglesias-Figueroa et al., 2018) with innate defence mechanism and exhibits a wide range of functional activities.The predominant function accomplished by Lf is facilitation of two proteins i.e. transferrin and ovotransferrin having metal binding sites which binds with ferric ion (Fe^{3+}) and bicarbonate ion.(HCO3-), respectively. It has great importance in stimulation of immune system leading to increased macrophage activity and decreased inflammatory cytokines such as TNF-α and nitric oxide. Several reports suggest dependence of lacoferrin's immunomodulatory mechanism on degree of its glycosylation (Zimecki et al., 2014). Actually, these are the main sites responsible for recognition of certain cell receptors of immune system as they remain exposed to external cell surface (Latorre et al., 2010). The ability of Lf for sequestering iron facilitates its antioxidant activity (Sharma et al., 2015), still, some animal studies have provided stronger evidence of Lf in treating cancer also (Gill & Cross, 2000). Antimicrobial activity possessed by Lf is broad spectrum, securing protective action against gram positive as well as gram negative bacteria, viruses, fungi, and parasites (Legrand et al., 2008). Lf have some role in glucose absorption especially under inflammatory conditions such as carcinogenesis, arthritis etc.whichusually inhibit intestinal prostaglandin E2 and thereby decreases Na^+ dependent glucose transportation. Hence, in such conditions Lf treatment can enhance glucose transport by increasing the SGLT1 activity.

Lysozyme, Lactoperoxidase and Casein Macropeptides

The lysozyme present in whey is both lysozyme c (found in hen egg white) and lysozyme g (found in goose egg white). It's antimicrobial activity against Gram-positive as well as Gram-positive due to bacterial cell wall lysis is well documented (Min et al., 2005). Lactoperoxidaseis a glycoprotein with 0.06-0.71% iron content. Lactoperoxidase shows the antimicrobial activity in presence of thiocyanate ion and hydrogen peroxide (Madureira et al., 2007). Casein macropeptides are derived by proteolysis of casein by proteinases. Glycomacropeptide (GMP) is obtained in whey due to enzymatic hydrolysis of 106-107 bond of κ-casein in bovine milk. It lacks aromatic amino acids but represents a significant source of branched chain amino acids.GMP have been reported to control lever disorders and shows opioid and satiety effect. GMP had been recommended for the supplementation of infant milk formulaedue to its positive effect on absorption of calcium, iron or zinc (Kelleher et al., 2003).

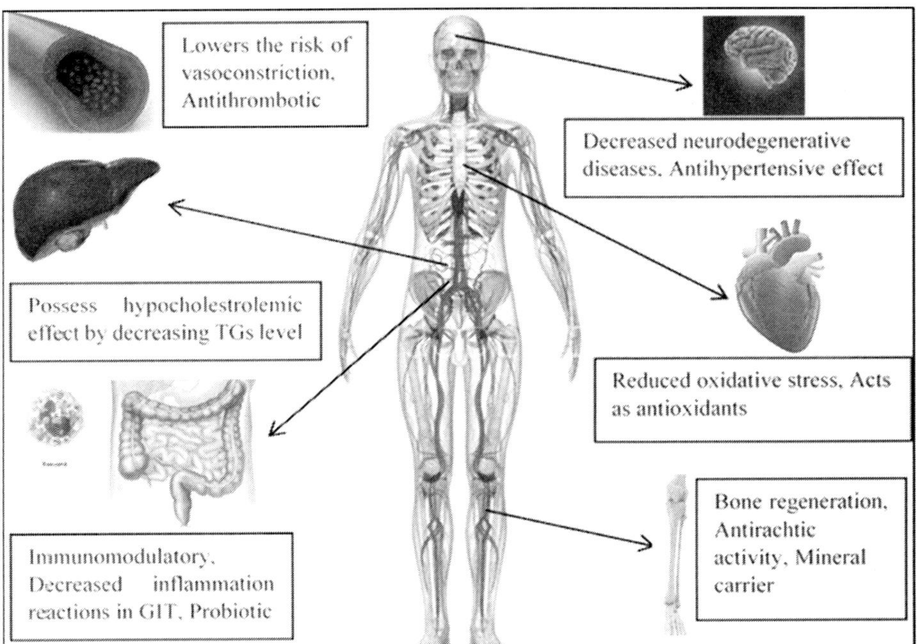

Fig. 1: Illustrative representation of beneficial effects of milk bioactive components on human body (TGs- Triglycerides, GIT- Gastrointestinal tract)

Milk Lipids

Fat is most variable component of milk with fluctuating fatty acid composition and amount of saturated and unsaturated fatty acids. Generally, eating healthy is synonymously attached with avoiding dietary fat, cholesterol and saturated fat because of their role in coronary heart disease (CHD). However, several components in milk fat such as: conjugated linoleic acid, omega-3 fatty acids, oleic acids, vaccenic acids, sphingolipids, butyric acid, ether lipids and 13-methyltetradecanoic acid have been documented for their health and nutritional benefits (Bauman et al., 2011). Milk fat is not only the source of bioactive components but also serves as carrier medium for nutrients, such as fat-soluble vitamins. Butyric acid, found only in milk fat, is anticarcinogenic and also forms a protective layer with etheric lipids, CLA and A, D vitamins against non-contagious diseases (Parodi, 2004). Caprylic, capric and lauric acids have been reported to have antiviral, antibacterial, anticaries and antiplaque activity (Thormar & Hilmarsson, 2007). The total milk fatty acid consists of 60-70% saturated fatty acid, which is precluded by most nutritionist for reducing cardiovascular diseases and plasma cholesterol reduction. Though, the most appreciated fact is that SFA vary in their structure and some of them have no effect on plasma cholesterol. As per a report only three SFA (lauric, myristic and palmitic) are associated with enhancement of blood cholesterol (German and Dillard, 2006). Despite of these reports, regulating bodies like USDA have recommended the intake of dietary saturated fatty acid to less than 10% of total calories (Hoenselaar, 2012).

Trans fatty acid (TFA) constitutes about 2-4% of milk fat, majorly trans-oleic acid and limited portion of trans-poly unsaturated fatty acid such as conjugated linoleic acid. Recent policies have aimed at labelling of trans fatty acid content on the packages because of their considerable adverse role in chronic and cardiovascular diseases. TFA fatty acid are produced either by chemical process i.e. industrial production of partially hydrogenated vegetable oil: PHVO) and during rumen biohydrogeneration. TFA consumption from PHVO source had reduced considerably while ruminant source TFA's intake is increasing owing to the significant amount of vaccenic acid (Gebauer et al., 2011). Approximately, thirty-three percent of the milk fatty acids are unsaturated (Oleic acid, conjugated linoleic acid and omega-3 fatty acid), which demonstrates cholesterol lowering tendency.

Omega-3 fatty acid (ω-3 FA) is important for growth, development and prevention of diseases like neurological disorders and other chronic diseases. α-linoleic acid (α-LA) is the major ω-3 FA in milk fat. ω-3 FA normally constitutes 0.5% of total milk fatty acid and a limiting enzyme, Δ6-desaturase is responsible for its conversion to long chain ω-3 FA (eicosapentaenoic acid, EPA and docosahexaenoic acid, DHA). α-linoleic acid conversion to EPA and DHA is essential for its beneficial effects i.e. prevention of chronic diseases (Whelan& Rust, 2006). An increased focus for the enhancement of ω-3 FA content of milk fat have led to approaches like feeding animals with flaxseed, linseed oil and pasture owing to their richness in α-LA. Though, due to limited activity of Δ6-desaturase in cow's mammary gland, this approach had not shown significant results (Palmquist et al., 2005). The addition of fish oil or its by-products to dairy cattle feed have led to generation of specialty milk with enhanced EPA and DHA content (Bauman et al., 2011).

Conjugated linoleic acids (CLA) were stamped as bioactive food components by Pariza and more than 20 different isomers of CLA have been identified in milk fat till date. The generation of CLA in milk fat is due to biohydrogenation of PUFA and rumenic acid represents the major CLA isomer in milk fat. The rumenic acid consideration as functional food component is due to its role in anti-cancer and anti-atherogenic effects (Lock et al., 2009). Several other CLA isomers are also present in milk fat but their low concentration laggards them to be considered as functional food component. However, a combined mixture of CLA isomers have shown significant activity against obesity and diabetes.

Milk Fat Globule Membrane as Potential Functional Ingredient

Milk contains many components that exhibit physiological functions. Among various constituents there is a thin membrane called, milk fat globule membrane (MFGM) which surrounds milk fat globules. This membrane, with a thickness of 10-20 nm, acts as an emulsifier and protects the fat globules from coalescence and enzymatic degradation. It can be isolated from rich source like buttermilk and butter serum (Ahn et al., 2011). The lipid fraction of MFGM are primarily polar lipids, consisting of phospholipids and sphingolipids which play a vital role in antiproliferative pathways associated with the suppression of colon carcinogenesis. Depending on the source, 25-70% of the MFGM consist of proteins such as butyrophilin which shows molecular mimicry with myelin oligodendrocyte glycoprotein (MOG). Xanthine oxidase has a beneficial role as an antimicrobial agent in gastrointestinal tract (Conway et al., 2014). Other components present in MFGM also has a specific technological and nutritional properties such as cholesterol lowering, prevention of multiple sclerosis and autism. Rich source of this membrane could potentially be incorporated in some products to enhance emulsion stability and to lower the interfacial tension (Malik et al., 2015). Hence, it is one of the most potent fractions that own numerous nutraceutical and functional properties.

Milk Carbohydrates

Carbohydrates are the most important source of energy or fuel for sports activities (>1.5 hours of exercise) as depletion of small reserves of liver and muscle glycogen leads to fatigue. The budding benefits of milk-based drinks as compared to commercial carbohydrate beverages during and before sports activities had been documented by Lee and co-workers (2008). Lactose is the major carbohydrate present in milk which is fermented to lactic acid by LAB, leading to solubilisation of proteins, minerals thereby enhancing calcium absorption, inhibition of harmful microorganisms and enhanced digestibility. Lactulose is a disaccharide formed during heat treatment of milk and has beneficial effect on human health by promoting the growth of probiotic microorganisms (Gibson, 2004).

Other minor carbohydrates like neutral and acidic oligosaccharides, glucose, fructose and glucosamine are also present in milk. Oligosaccharides (OS) consists mainly three to ten monosaccharide and structural and compositional variation have been reported in OS of different species (Mussatto & Mancilha, 2007). Human milk OS promotes the growth of bifidus flora and 3'-GL oligosaccharide presents in dromedary camel milk serves as prebiotic component (Alhaj et al., 2013). Human milk oligosaccharides have also been reported as potent inhibitors to adhesion of bacterial cell to epithelial surface, thus preventing bacterial infection (Kunz & Rudloff, 2006). The oligosaccharide present in milk possesses the bifidobacterium growth enhancing activity (Bezkorovainy & Topouzian, 1981). Some of the major health benefits of milk OS are:

- Promotes the growth of *Bifidobacterium* species in GI tract
- Inhibition of the growth of pathogenic microorganisms
- Reduction of type 2 diabetes and obesity
- Reduction of CVD, chronic diseases and diarrhoea
- Enhanced absorption of minerals

Milk and Dairy Product's Minerals and Vitamins as Functional Ingredients

Minerals in milk are found majorly in soluble phase and remains unchanged unless precipitated, however, vitamins on the other hand are either water (B and C) or fat soluble (A, D, E and K). Nevertheless, vitamins degrade with heat or light exposure. Milk contains 0.9% minerals which mainly consists of sodium, potassium, calcium, magnesium, chloride, phosphate and sulphate.

Calcium is the major mineral supplied by milk and regulates blood clotting, contraction and relaxation of muscles. Dairy ingredients, mainly whey is excellent source of calcium. Milk of different species are differing in majorly present mineral such as buffalo milk is somewhat higher in calcium and copper, while being lower in potassiumand zinc; sheep milk demonstrates enhanced calcium, phosphorus and zinc; goat milk is very similar to cow's milk; camel milk is lower in iron but higher in copper and zinc; mare milk is lower in phosphorus. Calcium, potassium and magnesium present in milk combat high blood pressure (Bornaz et al., 2009). Vitamin D along with calcium has been reported to prevent osteoporosis. Fat soluble vitamins (A, D, E and K), β-carotene (pro-vitamin A) have been found involved in brain development (Ebringer et al., 2008).

Milk Derived Bioactive Components from Metabolic Activities

Bioactive peptides (BAP) are specific protein fragments that are produced *in vivo* or *in vitro* by enzymatic hydrolysis of food proteins which exerts positive health impacts beyond basic nutrition.These health benefits are exerted through binding to specific receptors thereby inducing physiological responses (Fitz Gerald & Murray, 2006). Milk is considered as the most important source of bioactive peptides as they are secreted in active form by mammary gland (Meisel, 2004). As per a study by Sipola et al. (2001) milk is found to be a better carrier of these BAP as compared to feeding them with pure water because of more bioactivity effect in milk. Epidermal growth factor (EGF), transforming growth factor (TGF),nerve growth factor (NGF), insulin and insulin like growth factors I and II (IGF- I and IGF- II) are naturally available bioactive peptides in bovine milk. Bioactive peptides can also be produced by proteolysis of individual casein and whey proteins (Hamers-Casterman et al., 1993). These bioactive peptides can be produced during ripening of cheeses by microbial proteolysis thus creating new market segment of 'naturally occurring functional foods'. The range of bioactive peptides varies with the proteolytic activity of microorganisms, amino acid composition and sequence of peptide. As a part of microorganism's metabolic activity, hydrolysis of protein occurs and peptides are released.Total enzymatic synthesis and recombinant DNA technology are the approaches available for peptide synthesis, with later being expensive in research and development phase. However, once the technology is established large quantity of BAP could be obtained through fermentation (Gill et al., 1996).

Commercial Products Based on Functional Dairy Ingredients

Functional ingredients in milk have attracted attention as suitable ingredients for health promoting foods. Such formulations are commercially available in various forms such as milk-based beverages, fermented drinks, powder and capsules (Table 2). "Festivo" cheese produced using cheese starter in combination with *Lactobacillus acidophilus* and *Bifidobacterium*had been reported to demonstrate increase in concentration of ACE inhibitory activity during cheese ripening and decrease when proteolysis exceeds a certain limit (Ryhanen et al., 2001).

Table 2: Functional dairy products currently available in market

Product name	Description	Brand
Calpis	Sour milk	Calpis Co., Ltd.
Yakult®	Probiotic drink	Danone
DanActive®	Probiotic drink	Danone
Kefir	Fermented milk with kefir grains	Lifeway
Festivo	Low fat hard cheese	MTT Agrifood, Finland
Ultimate Greek®	Greek yoghurt	Danone
Yoghurt squishy	Flavoured yoghurt drink	Brooklea®
Epigamia	Greek yoghurt	Drums Food International®
C12	Hydrolysate	DMV International, Netherlands
Fruit yoghurt	Yoghurt with fruits pieces	Mother Dairy®
ProLife	Probiotic ice cream	Amul®
Advanced dahi	Probiotic *dahi*	Mother Dairy®
Capolac	Milk Mineral Concentrate	Arla Foods Ingredients, Sweden
Make your own yoghurt	Yoghurt powder	Zoh Probiotics®

Challenges in Development of Functional Dairy Foods in India

In India, we have traditional products touted as functional but have little scientific validation. Regulation will thus have to evolve to promote R&D, ensure validation and prevent exploitation of consumer Companies will also have to be sincere and honest in their claims while marketing and communicating with consumers till appropriate regulations for scientific validation are evolved. Processors will need l provide an optimal merger between taste, convenience and health attributes. Companies will require expert knowledge in flavour masking fortification know-how and delivery systems.

Technological Challenges

Four different technological hurdles have to be overcome before a product containing bioactive substances is ready to consider marketing:

- Isolation of the desired components,
- Pre-establishment of the biological activity,
- Incorporation of the bioactive components into a formulated product,
- Verification of efficacy and safety of final product.

Such a sequence of experimental events is also required for the introduction of new food additive additional is a thorough proof of the claimed benefit of the bioactive components. This applies especially when the bioactive component is a completely new substance never consumed before in significant amounts. Separation, purification and production at industrial level of such nutraceuticals must be thought in terms of integrated and high added value. Membrane technologies (MF, UF, NF) provide key opportunities to manufacture milk nutraceuticals in native state. The incorporation of bioactive components into processed foods, its delivery and bioavailability are other important issues that need reprisal.

Scientific Validation of Functional Foods

The scientific evidence for functional foods and their physiologically active components can be categorized into four distinct areas: (a) clinical trials, (b) animal studies, (c) experimental in vitro laboratory studies, and (d) epidemiologic studies. Much of the current evidence for functional foods lack well-designed clinical trials; however, the foundational evidence provided through the other types of scientific investigation is substantial for several of the functional foods and their health-promoting components.

Safety Issues

Although increasing the availability of healthful foods including functional foods in the diet is critical to ensuring a healthier population, safety is a critical issue. The optima: levels of the majority of biologically active components currently under investigation have yet to be determined. The benefits and risks to individuals and populations as a whole must be weighed carefully when considering the widespread use of physiologically-active functional foods. Knowledge of toxicity of functional foe components is crucial to decrease the risk: benefit ratio. The safety issues related to probiotic microorganisms that has to be considered include

- intrinsic properties of the probiotic strains
- pharmacokinetics of probiotic strains
- interaction between probiotic strains and the host

- Knowledge of toxicity of functional food components is crucial to decrease the risk: benefit ratio

Conclusion

Consumer interest in the relationship between diet and health has increased the demand for information on functional foods. Rapid advances in science and technology, increasing healthcare costs, changes food laws affecting label and product claims, an aging population, and rising interest in attaining wellness through diet are among the factors fuelling interest in functional foods. Credible scientific research indicates many potential health benefits from milk components. It should be stressed, however, that functional foods are not a magic bullet or universal panacea for poor health habits. There are no "good" or "bad" foods but there are "good" or "bad" diets. Research into functional foods will not advance public health unless the benefits of the foods are effectively communicated to the consumer. The communication guidelines need to be established to ensure that research results about nutrition food safety and health are communicated in a clear, balanced and non-misleading manner.

References

Ahn, Y. J., Ganesan, P., Kwak, H. S. (2011). Composition, structure, and bioactive components in milk fat globule membrane.Korean Journal for Food Science of Animal Resources, 31 (1), 1-8.

Alhaj, O.A., Taufik, E., Handa, Y., Fukuda, K., Saito, T., Urashima, T. (2013). Chemical characterization of oligosaccharides in commercially pasteurized dromedary camel (camelus dromedarius) milk. International Dairy Journal, 28, 70–75.

Bauman, D.E., Harvatine, K.J., Lock, A.L. (2011). Nutrigenomics, rumen-derived bioactive fatty acids, and the regulation of milk fat synthesis. Annual Review of Nutrition, 31, 299–319.

Bezkorovainy, A., Topouzian, N. (1981). Bifidobacterium bifidus var. pennsylvanicus growth promoting activity of human milk casein and its derivatives.International Journal of Biochemistry,.13 (5), 585-590.

Bornaz, S., Sahli, A., Attalah, A., Attia, H. (2009). Physicochemical characteristics and renneting properties of camels' milk: a comparison with goats', ewes', and cows' milks. International Journal of Dairy Technology, 62 (4), 505–513.

Cakebread, J. A., Humphrey, R., Hodgkinson, A. J. (2015). Immunoglobulin A in bovine milk: a potential functional food? Journal of agricultural and food chemistry,.63 (33), 7311-7316.

Cho, Y., Batt, C. A., Sawyer, L. (1994). Probing the retinol-binding site of bovine beta-lactoglobulin.Journal of Biological Chemistry,.269 (15), 11102-11107.

Conway, V., Gauthier, S. F., Pouliot, Y. (2014). Buttermilk: much more than a source of milk phospholipids.Animal Frontiers,.4 (2), 44-51.

Ebringer, L., Ferencík, M., Krajcovic, J. (2008). Beneficial health effects of milk and fermented dairy products.Folia Microbiologica,.53 (5), 378-394.

Fiat, A.-M., Migliore, D. Jollès, P. (1993). Biologically active peptides from milk proteins with emphasis on two examples concerning antithrombotic and immune stimulating activities. Journal of Dairy Science, 76 (1), 301–310.

FitzGerald, R.J. Murray, B.A. (2006). Bioactive peptides and lactic fermentations. International Journal of Dairy Technology, 59, 118–125.

Ganjam, L. S., Thornton Jr, W. H., Marshall, R. T., & MacDonald, R. S. (1997). Antiproliferative effects of yogurt fractions obtained by membrane dialysis on cultured mammalian intestinal cells.Journal of Dairy Science,.80 (10), 2325-2329.

Gauthier, S. F., Pouliot, Y., Saint-Sauveur, D. (2006). Immunomodulatory peptides obtained by the enzymatic hydrolysis of whey proteins.International dairy journal, 16 (11), 1315-1323.

Gebauer, S.K., Chardigny, J. M., Jakobsen,M.U., Lamarche, B., Lock, A.L., Proctor, S.D. Baer, D.J. (2011). Effects of ruminant trans fatty acids on cardiovascular disease and cancer: a comprehensive review of epidemiological, clinical, and mechanistic studies. Advances in Nutrition, 2, 332–354.

German, J.B., Dillard, C.J. (2006). Composition, structure and absorbtion of milk lipids: A source of energy, fat-soluble nutrients and bioactive molecules. Critical Reviews in food science and nutrition, 46:57-92.

Gibson, G.R. (2004). From probiotics to prebiotics and a healthy digestive system. Journal of Food Science. 69:M141-M143.

Gill, H. S., Cross, M. L. (2000). Anticancer properties of bovine milk.British Journal of Nutrition,.84(S1), 161-166.

Gill, I., López-Fandiño, R., Jorba, X. Vulfson, E.N. (1996). Biologically active peptides and enzymatic approaches to their production. Enzyme Microbial Technology, 18, 162–183.

Hakansson, A., Zhivotovsky, B., Orrenius, S., Sabharwal, H., &Svanborg, C. (1995). Apoptosis induced by a human milk protein.Proceedings of the National Academy of Sciences,.92(17), 8064-8068.

Hamers-Casterman, C., Atarouch, T.,Muyldermans, S., Bendolman, N. Hamers, R. (1993). Naturally occurring antibodies devoid of light chains. Nature, 363, 446–448.

Heaney, R. P., Saito, Y., Orimo, H. (1994). Effect of caseinphosphopeptide on absorbability of co-ingested calcium in normal postmenopausal women.Journal of Bone and Mineral Metabolism,.12(1), 77-81.

Hoenselaar, R. (2012). Saturated fat and cardiovascular disease: the discrepancy between the scientific literature and dietary advice. Nutrition, 28, 118–123.

Kapila, S., Jian, S., Haque, E., Sabhikhi, L., Dang, A.K. (2008). Immunoregulatory response of Lactobacillus helveticus fermented milk in mouse. Milchwissenschaft, 63(4), 367-349.

Kunz, C. Rudloff, S. (2006). Health promoting aspects of milk oligosaccharides. International Dairy Journal, 16, 1341–1346.

Latorre, D., Puddu, P., Valenti, P., Gessani, S. (2010). Reciprocal interactions between lactoferrin and bacterial endotoxins and their role in the regulation of the immune response.Toxins,.2 (1), 54-68.

Lee, J.K., Maughan, R.J., Shirreffs, S.M., Watson, P. (2008). Effects of milk ingestion on prolonged exercise capacity in young, healthy men. Nutrition, 24(4), 340–347.

Legrand, D., Pierce, A., Elass, E., Carpentier, M., Mariller, C., Mazurier, J. (2008). Lactoferrin structure and functions. In.Bioactive components of milk.(pp. 163-194). Springer, New York, NY.

Lock, A.L., Kraft, J., Rice, B.H., Bauman, D.E. (2009). Biosynthesis and biological activity of rumenic acid: a natural CLA isomer. In: Trans Fatty Acid in Human Nutrition, 2nd Edition (eds Destaillats, F., Sébédio, J.L., Dionisi, F. &Chardigny, J.M.), pp. 195–230. Bridgewater, UK: The Oily Press.

Loimaranta, V., Laine, M., Soèderling, E., Vasara, E., Rokka, S., Marnila, P., &Tenovuo, J. (1999). Effects of bovine immune and non-immune whey preparations on the composition and pH response of human dental plaque.European journal of oral sciences,.107(4), 244-250.

Malik, P., Danthine, S., Paul, A., Blecker, C. (2015). Physical-chemical properties of milk fat globule membrane at different stages of isolation.Scientific Bulletin. Series F, Biotechnologies,.19, 154-159.

Markus, C. R., Olivier, B., de Haan, E. H. (2002). Whey protein rich in α-lactalbumin increases the ratio of plasma tryptophan to the sum of the other large neutral amino acids and improves cognitive performance in stress-vulnerable subjects.The American journal of clinical nutrition,.75(6), 1051-1056.

Meisel, H. (2004). Multifunctional peptides encrypted in milk proteins. BioFactors, 21, 55–61.

Meisel, H., FitzGerald, R. J. (2003). Biofunctional peptides from milk proteins: mineral binding and cytomodulatory effects. Current pharmaceutical design, 9(16), 1289-1296.

Mellander, O. L. O. F. (1950). The physiological importance of the casein phosphopeptide calcium salts. II. Peroral calcium dosage of infants. Some aspects of the pathogenesis of rickets.Acta SocietatisBotanicorumPoloniae,.55, 247-257.

Mills, S., Ross, R. P., Hill, C., Fitzgerald, G. F., Stanton, C. (2011). Milk intelligence: Mining milk for bioactive substances associated with human health.International dairy journal,.21(6), 377-401.

Min, S., Harris, L.J., Han, J.H. Krochta, J.M. (2005). Listeria monocytogenes inhibition by whey protein films and coatings incorporating lysozyme. Journal of Food Protection, 68(11), 2317–2325.

Mussatto, S.I. Mancilha, I.M. (2007). Non-digestible oligosaccharide: A review. Carbohydrate Polymers, 68, 587–597.

Nagaoka, S., Futamura, Y., Miwa, K., Awano, T., Yamauchi, K., Kanamaru, Y., Kuwata, T. (2001). Identification of novel hypocholesterolemic peptides derived from bovine milk β-lactoglobulin.Biochemical and Biophysical Research Communications, 281(1), 11-17.

Nakamura, Y., Yamamoto, N., Sakai, K. & Takano, T. (1995). Antihypertensive effect of sour milk and peptides isolated from it that are inhibitors to angiotensin I-converting enzyme. Journal of Dairy Science, 78, 1253–1257.

Palmquist, D.L., Lock, A.L., Shingfield, K.J. Bauman, D.E. (2005). Biosynthesis of conjugated linoleic acid in ruminants and humans. Adances in Food and Nutrition Research, 50, 179–218.

Park, Y. W. (Ed.). (2009).Bioactive components in milk and dairy products. John Wiley & Sons.

Parodi, P.W. (2004). Milk in human nutrition. Australian Journal of Dairy Technology. 59:3-59.

Qiang, X., YongLie, C. QianBing, W. (2009). Health benefit application of functional oligosaccharides. Carbohydrate Polymers, 77, 435–441.

Regester, G. O., Smithers, G. W., Mitchell, I. R., McIntosh, G. H., & Dionysius, D. A. (1997). Bioactive factors in milk: natural and induced.

Rokka, T., Syväoja, E. L., Tuominen, J., & Korhonen, H. J. (1997). Release of bioactive peptides by enzymatic proteolysis of Lactobacillus GG fermented UHT-milk.

Ryhanen, E., Leppala, A.P. Pahkala, E. (2001). A new type of ripened, low fat cheese with bioactive properties. International Dairy Journal, 11:441-447.

Sharma, R., Chakraborty, D., Gupta, P. (2015). Bovine lactoferrin and its functions in animals-A review.Agricultural Reviews, 36(4).

Shinoda, I., Takase, M., Fukuwatari, Y., Shimamura, S., Köller, M., König, W. (1996). Effects of lactoferrin and lactoferricin® on the release of interleukin 8 from human polymorphonuclear leukocytes.Bioscience, biotechnology, and biochemistry, 60(3), 521-523.

Sipola, M., Finckenberg, P., Santisteban, J., Korpela, R., Vapaatalo, H. Nurminen, M.-L. (2001). Long term intake of milk peptides attenuates development of hypertension in spontaneously hypertensive rats. Journal of Physiology and Pharmacology, 52, 745–754.

Stanton, C., Ross, R. P., Fitzgerald, G. F., & Van Sinderen, D. (2005). Fermented functional foods based on probiotics and their biogenic metabolites.Current opinion in biotechnology, 16 (2):198-203.

Sternhagen, L. G., Allen, J. C. (2001). Growth rates of a human colon adenocarcinoma cell line are regulated by the milk protein alpha-lactalbumin. In Bioactive Components of Human Milk (pp. 115-120). Springer, Boston, MA.

Talukder, J. R., Griffin, A., Jaima, A., Boyd, B., Wright, J. (2014). Corrigendum: Lactoferrin ameliorates prostaglandin E2-mediated inhibition of Na+-glucose cotransport in enterocytes. Canadian journal of physiology and pharmacology, 92 (6), 519-519.

Thormar, H Hilmarsson, H. (2007). The role of microbicidal lipids in host defence against pathogens and their potential as therapeutic agents. Chem. Phys. Lipids, 150:1-11.

Wang, Y.W. Jones, P.J. (2004). Conjugated linoleic acid and obesity control: efficacy and mechanisms. International Journal of Obesity and Related Metabolic Disorders, 28, 941–955.

Whelan, J. Rust, C. (2006). Innovative dietary sources of n-3 fatty acids. Annual Review of Nutrition, 26, 75–103.

Wu, S. Y., Perez, M. D., Puyol, P., Sawyer, L. (1999). β-Lactoglobulin binds palmitate within its central cavity. Journal of Biological Chemistry, 274 (1), 170-174.

Xu, R.J. (1998). Bioactive peptides in milk and their biological and health implications. Food Reviews International, 14 (1), 1–16.

Zimecki, M., Artym, J., Kocięba, M., Duk, M., &Kruzel, M. L. (2014). The effect of carbohydrate moiety structure on the immunoregulatory activity of lactoferrin in vitro. Cellular & molecular biology letters, 19(2), 284.

13

Probiotics, Prebiotics and Synbiotics The Dairy Functional Food

[1]Devenda Jain, [1]Heena Saheewala, [1]Suman Sanadhya [1]Ramandeep Kour* and *[1]Arunabh Joshi

Department of Molecular Biology and Biotechnology
Rajasthan College of Agriculture
Maharana Pratap University of Agriculture and Technology Udaipur, Rajasthan

Introduction

During the first 50 years of the 20th century, scientific focus was on the identification of essential elements and their role in the prevention of various dietary deficiency diseases. This emphasis on nutrient deficiencies or "undernutrition" shifted dramatically, however, during the 1970s when diseases linked to excess and "overnutrition" became a major public health concern. "Let food be thy medicine and medicine be thy food", the age old quote by Hippocrates is the ideology of today's health conscious population. Eli Metchnikoff, the Russian Nobel prize winner was the first one to recognize the beneficial role of selected bacteria on gastrointestinal tract of humans. Subsequently the "Theory of Longevity" by Metchnikoff was correlated with prolonged youth and a healthy old age, observed largely in Balkan peasants of those times, who used cultured milks in their diet (Kaufmann, 2008). Since then, the quest continues for understanding the role of wide range of food components and nutrients in enhancing health or preventing chronic diseases. The research in this field has resulted in a plethora of new labels for pro-, pre- and synbiotics.

Probiotics

The term probiotic comes from the Latin or Greek meaning pro, "before, forward", and bios, or "life" thus probiotics are life-promoting. In this case, we use the term probiotics to refer to beneficial bacteria. Bulgarian yoghurt (sour milk), commonly consumed plain, is popular for its taste, aroma, and quality. The qualities arise from the *Lactobacillus bulgaricus* and *Streptococcus thermophilus* culture strains used in Bulgaria. In 1907, Metchnikoff proposed that the acid producing bacteria in fermented milk products could prevent "fouling" in the large intestine and, if consumed regularly, lead to a longer, healthier life. In early 1930's, in Japan, Minoru Shirota developed a fermented milk product called Yakult and in 1935 he started marketing Yakult as a probiotic yogurt-like product made by fermenting a mixture of skimmed milk with a special strain of *Lactobacillus casei shirota.*

According to the currently adopted definition by FAO/WHO, probiotics are: "Live microorganisms which when administered in adequate amounts confer a health benefit on the host". Lactic acid bacteria (LAB) and *bifidobacteria* are the most common types of microbes used as probiotics; but certain yeasts and bacilli may also be used. These are a part of the intestinal tract that helps maintaining the natural balance of organisms (microflora) in the intestinal camera. The normal human digestive tract contains about 400 types of probiotic bacteria that reduce the growth of harmful bacteria and promote a healthy digestive system. The largest group of probiotic bacteria in the intestine is lactic acid bacteria, of which *Lactobacillus acidophilus*, found in yogurt with live cultures, is the best known. Yeast is also a probiotic substance and available as dietary supplements. Thus, Probiotic microorganisms are able to survive the passage through the digestive system by colonizing intestinal epithelia utilizing the nutrients and substrates in a normal diet and exerting a beneficial effect on the host. All probiotics work effectively, if taken after meals, when stomach acid is lowest. Stomach acid destroys up to 99.9% of probiotics, if taken before a meal but only about 90%, if taken after meal. Thus, after meal intake ensures maximum delivery to the small intestine, where they reproduce rapidly, populating the colon. They do most of their good work in the small and large intestines, but they also help the mouth and esophagus to remain uninfected.

Some of the popularly used probiotic microorganisms are *Lactobacillus rhamnosus*, *Lactobacillus reuteri, Bifidobacteria* and certain strains of *Lactobacillus casei, Lactobacillus acidophilus-group, Bacillus coagulans, Escherichia coli* strain Nissle 1917, certain enterococci, especially *Enterococcus faecium* SF68, and the yeast *Saccharomyces boulardii*. Bacterial spore formers, mostly of the genus *Bacillus* dominate the scene. These probiotics are added to foods, particularly fermented milk products, either singly or in combinations. New genera and strains of probiotics are continuously emerging with more advanced and focused research efforts. Probiotic products may contain either a single strain or a mixture of two or more strains. E.g. #VSL3 is a mixture of eight different probiotic strains. Probiotic effects are very strain specific and cannot be generalized. A single strain may exhibit different benefits when used individually and in combination.

FAO and WHO have jointly put forward guidelines in order to set out a systematic approach for an effective evaluation of probiotics in foods to substantiate the health claims and benefits and the guidelines to be used as global standard for evaluating probiotics in food that could result in the substantiation of health claims are:

- Strain identification.
- Functional characterization of the strain(s) for safety and probiotic attributes.
- Validation of health benefits in human studies.
- Honest, not misleading labelling of efficacy claims and content for the entire shelf life.

Prebiotics

A prebiotic is a non-digestible component which beneficially affects the host by selectively stimulating the growth and/or activity of one or a limited number of colonic bacteria, thereby FAO/WHO defines prebiotics as a non-viable food component that confer health benefit(s) on the host associated with modulation of the microbiota. Prebiotics form a group of diverse carbohydrate ingredients that are poorly understood with reference to their origin, fermentation profiles, and dosages required for health effects. Some of the sources of prebiotics include: breast milk, soybeans, inulin sources (like Jerusalem artichoke, chicory roots etc.), raw oats, unrefined wheat, unrefined barley, yacon, non-digestible carbohydrates, and in particular non-digestible oligosaccharides. An ideal prebiotic should be

- Resistant to the actions of acids in the stomach, bile salts and other hydrolyzing enzymes in the intestine.
- Should not be absorbed in the upper gastrointestinal tract.
- Be easily fermentable by the beneficial intestinal microflora (Kuo 2013).

Prebiotics like inulin and pectin exhibit several health benefits like Reducing the prevalence and duration of diarrhea, relief from inflammation and other symptoms associated with intestinal bowel disorder and protective effects to prevent colon cancer (Pena 2007). They are also implicated in enhancing the bioavailability and uptake of minerals, lowering of some risk factors of cardiovascular disease, and promoting satiety and weight loss thus preventing obesity (Pokusaeva et al.,2011).

Prebiotics may be used as an alternative to probiotics or as an additional support for them. Long-term stability during the shelf-life of food, drinks, and feed, resistance to processing, and physical and chemical properties that exhibit a positive effect on the flavour and consistence of products may promote prebiotics as a competition to probiotics. Additionally, resistance to acids, proteases, and bile salts present in the gastrointestinal tract may be considered as other favourable properties of prebiotics. Prebiotic substances selectively stimulate microorganisms present in the host's intestinal ecosystem, thus eliminating the need for competition with bacteria. Stimulation of the intestinal microbiota by prebiotics determines their fermentation activity, simultaneously influencing the SCFA level, which confers a health benefit on the host . Moreover, prebiotics cause a reduction of intestinal pH and maintain the osmotic retention of water in the bowel. However, it should be considered that an overdose of prebiotics may lead to flatulence and diarrhoea—these effects are absent in the case of excessive consumption of probiotics. Prebiotics may be consumed on a long-term basis and for prophylactic purposes. Moreover, when used at correct doses, they do not stimulate any adverse effects, such as diarrhoea, susceptibility to UV light or hepatic injuries caused by antibiotics. Prebiotic substances are not allergenic and do not proliferate the abundance of antibiotic resistance genes. Of course, the effect of the elimination of selected pathogens achieved by the use of prebiotics may be inferior to antibiotics, but the properties mentioned above make them a natural substitute for antibiotics.

Synbiotics

Synbiotics are a combination of prebiotics and probiotics that are believed to have a synergistic effect by inhibiting the growth of pathogenic bacteria and enhancing the growth of beneficial organisms. The concept has been proposed to characterize health-enhancing foods and supplements used as functional food ingredients in humans (Gibson, 2004). Synbiotics were developed to overcome possible survival difficulties for probiotics. It appears that the rationale to use synbiotics, is based on observations showing the improvement of survival of the probiotic bacteria during the passage through the upper intestinal tract. The probiotic strains used in synbiotic formulations include *Lacbobacilli, Bifidobacteria* spp, *S. boulardii, B. coagulans* etc., while the major prebiotics used comprise of oligosaccharides like fructooligosaccharide (FOS), GOS and xyloseoligosaccharide (XOS), inulin, prebiotics from natural sources like chicory and yacon roots, etc. The health benefits claimed by synbiotics consumption by humans include:

- Increased levels of lactobacilli and bifidobacteria and balanced gut microbiota.
- Improvement of liver function in cirrhotic patients.
- Improvement of immunomodulating ability.
- Prevention of bacterial translocation and reduced incidences of nosocomial infections in surgical patients, etc. (Zhang *et al.*, 2010).

Health Benefits of Probiotics, Prebiotics and Synbiotics

The most important and documented beneficial effects of probiotics include the prevention of diarrhea, constipation, changes in bile salt conjugation, enhancement of antibacterial activity, anti-inflammatory. Furthermore, they also contribute to the synthesis of nutrients and improve their bioavailability; some probiotics are known to exert anti-oxidative activity in the form of intact cells or lysates. They may be helpful in the treatment of inflammatory enteral conditions, including ulcerative colitis, Crohn's disease, and non-specific ileitis. A positive effect on the urogenital system (prevention and treatment of Urinary Tract Infections (UTIs) and bacterial vaginitis) constitutes an excellent example of the benefits associated with the use of probiotics (Pandey, 2015). There were attempts to apply probiotics to pregnant women and neonates in order to prevent allergic diseases such as atopic dermatitis. There is evidence that the consumption of probiotics-containing dairy products results in the reduction of blood cholesterol, which may be helpful in the prevention of obesity, diabetes, cardiovascular diseases, and cerebral stroke. The application of *Saccharomyces boulardii* yeast to patients with acute, watery diarrhoea resulted in the cure and reduced frequency of that type of complaints in two subsequent months. The efficacy of probiotic strains in the therapy of nosocomial, non-nosocomial, and viral diarrhoeas has also been documented. It turns out that probiotics may increase the amount of IgA antibodies, which leads to the arrest of a viral infection. Antibiotic-associated diarrhoea (AAD) is a common complication of most antibiotics and *Clostridium difficile* disease (CDD), which also is incited by antibiotics, and is a leading cause of nosocomial outbreaks of diarrhoea and colitis. Using meta-analyses, three types of probiotics (*Saccharomyces boulardii*, *Lactobacillus rhamnosus* GG, and probiotic mixtures) significantly reduced the development of antibiotic-associated diarrhea (Markowiak et al., 2017).

Among the advantages of prebiotics, reduction of the blood LDL (low-density lipoprotein) level, stimulation of the immunological system, increased absorbability of calcium, maintenance of correct intestinal pH value, low caloric value, and alleviation of symptoms of peptic ulcers and vaginal mycosis are important to mention. Other effects of inulin and oligofructose on human health are: the prevention of carcinogenesis, as well as the support of lactose intolerance or dental caries treatment.

The consumption of Synbiotics imparts beneficial effects on humans such as increased *Lactobacillus* and *Bifidobacterium* genus count and maintenance of balance of the intestinal microbiota; improved hepatic function in patients suffering from cirrhosis; improved immunomodulative abilities; and prevention of bacterial translocation and reduced incidence of nosocomial infections in patients' post-surgical procedures and similar interventions.

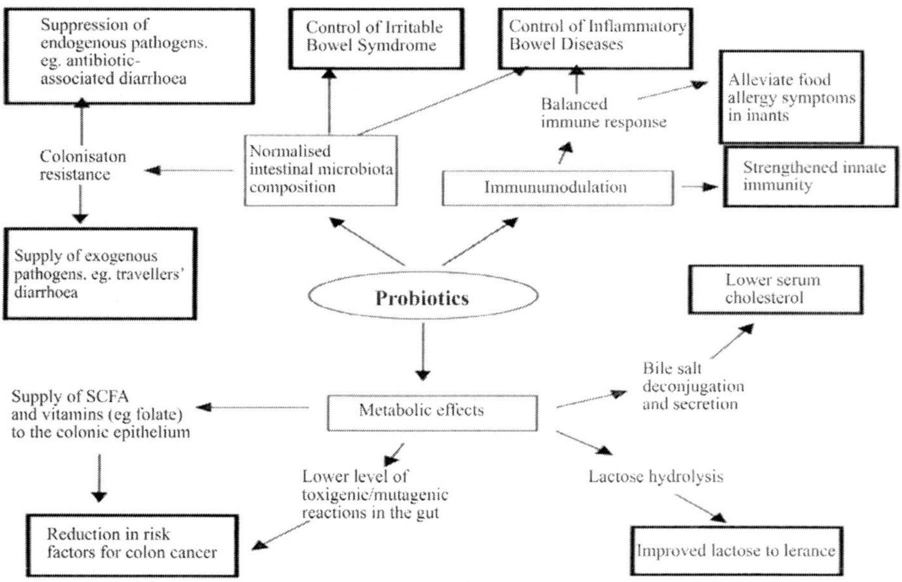

Proposed health benefits stemming from probiotic consumption.

Future Prospects

Concerted research efforts are being directed to establish the probiotic effects on cardiovascular disorders like myocardial infarction, atherosclerosis etc. (Loscalzo 2011). Neuro-gastroenterologist Dr. Gershon's working hypothesis postulates the existence of an enteric nervous system, its role and its participation in gut's physiology and other associated gut disorders (Gershon 1998). The afore mentioned hypothesis can be addressed by understanding the role of "Microbial endocrinology"- Probiotics synthesize as well as respond to the neuroactive compounds (Roshchina 2010).

Probiotics Market Share

According to the report, the global probiotics market was valued at approximately USD 40.09 billion in 2017 and is expected to generate revenue of around USD 65.87 billion by end of 2024, growing at a CAGR of around 7.35% between 2018 and 2024. Probiotic ingredients are sold in the form of capsules, tablets, powders, and gel. The major factor boosting the market growth is increasing consumer preference for natural products. Increasing concerns of consumers on preventive healthcare and effectiveness of probiotic bacteria on health drive the market growth. The demand of probiotics has increased due to upsurge in consumption of functional foods, which have potentially positive effects on health apart from providing basic nutrition. Apart from this, factors such as ban on the usage of antibiotics in animal feed in several European and North American countries, increasing product innovations in the form of chocolates, biscuits & others, increasing demand of probiotic based medicines due to proven positive results for reducing risk of chronic diseases such as diarrhoea & bowel diseases related to antibiotic use and rising demand for fermented food in daily diet. Additionally, increasing disposable income and improved standards of living in developing economies supplement the market growth.

Top Factors Impacting the World Probiotics Market

- Growing health concerns
- Low awareness among consumers
- Increasing investment on R&D

Conclusion

Probiotics, probiotics and synbiotics have systemic effects on the host's health metabolism and immune system. Utilization of prebiotics by probiotics should be a pre-requisite for symbiotic selection, in order to maintain a good synergy between the two and maximize the beneficial effects. By establishing the underlying mechanisms of probiosis and prebiosis, scientists would be able to design enhanced functional foods to improve host health. The ability to regulate the composition of the microbiota by prebiotic dietary substances and probiotic microorganisms is an interesting approach in the control and treatment of some major diseases. The recent advances in technology have enabled the deep sequencing and analysis of the unexpected diversity of the microorganisms in the GIT and it should be able to prevent the diseases and also facilitate to maintain a better health.

References

Gershon MD (1998) The second brain (pp. 4–7). Harper Collins Publishers.

Gibson GR, Probert HM, Loo JV, RastalL RA., & Roberfroid MB, (2004). Dietary modulation of the human colonic microbiota: updating the concept of prebiotics. Nutri Res Rev, 17(2), 259–275.

Kaufmann SH (2008) Immunology's foundation: the 100-year anniversary of the Nobel Prize to Paul Ehrlich and Elie Metchnikoff. Nat Immunol 9(7):705–712.

Kuo SM (2013) The interplay between fiber and the intestinal microbiome in the inflammatory response. Adv Nutr: Intern Rev J 4(1):16–28. doi:10.3945/an.112.003046

Loscalzo J (2011) Lipid metabolism by gut microbes and atherosclerosis. Circ Res 109 (2): 127-129.

Markowiak P and Katarzyna S (2017) Effects of Probiotics, Prebiotics, and Synbiotics on Human Health. Nutrients, 9, 1021; doi:10.3390/nu9091021.

Pandey KR, Naik SR, Vakil BV (2015) Probiotics, prebiotics and synbiotics- a review. J Food Sci Technol 52(12):7577–7587.

Peña AS (2007) Intestinal flora, probiotics, prebiotics, synbiotics and novel foods. Rev Esp Enferm Dig 99(11):653.

Pokusaeva K, Fitzgerald GF, van Sinderen D (2011) Carbohydrate metabolism in Bifidobacteria. Gen Nutr 6(3):285–306. doi:10.1007/s12263-010-0206-6.

Roshchina VV (2010) Evolutionary considerations of neurotransmitters in microbial, plant, and animal cells. in Microbial Endocrinol 17–52,Springer New York.

Zhang MM, Cheng JQ, Lu YR, Yi ZH, Yang P, Wu XT (2010) Use of pre-, pro-and synbiotics in patients with acute pancreatitis: a metaanalysis. World J Gastroenterol: WJG 16(31):3970. doi:10.3748/wjg.v16.i31.3970.

14

Microbial Functional Food Ingredients

Pramila Epparti, Nidhi Sori, Avinash P Sattur and Mahejibin Khan

Microbiology and Fermentation Technology, Department CSIR-Central Food Technological Research Institute, Mysore, Karnataka

Introduction

Hippocrate, an ancient physician, a great thinker and also often referred as father of medicine, for the first time conceive the idea nearly 2,500 years ago that a disease is caused due to natural factors such as environmental factors, diet, and living habits. He advocated the secret of healthy living in healthy food in his famous quote "Let food be thy medicine and medicine be thy food". But unfortunately, this wisdom about food has faded with the time and we became dependent on synthetic medicines and drugs for even life style disorders. In the last century, research advancement in the field of food science and nutrition established food not only as a source of nutrition but also a source of active food components that aid specific functions and rejuvenated the Hippocrates concept. According to the Department of Health and Human Services, USA, daily diet of a person, plays a crucial role in 5 of 10 diseases which are the leading causes of death, such as diabetes (noninsulin dependent or type 2), stroke, coronary heart disease (CHD), specific type of cancer, and atherosclerosis. In the mid-1980s, term "functional food" was introduce in Japan and defined as food that contain active components either from plant or animal origin and provide physiological benefits in addition to the basic nutritional needs.

Though, in general terms, almost all the foods are functional food to certain extent, as they provide protein, carbohydrates for energy or vitamins and minerals for cell function and deliver one or another physiological benefits (Hesler, 2002). But to be more precise, different scientific agencies and food associations defined functional food in different term as, "Any modified food or food ingredient that may provide a health benefit beyond the traditional nutrients it contains" by The National Academy of Sciences' Food and Nutrition Board In 1994)(Committee report 1994). "Foods that, by virtue of the presence of physiologically-active components, provide a health benefit beyond basic nutrition" (The International Life Sciences Institute, 1999). American Dietetic Association(1999) states that whole food, enriched, fortified, or enhanced food, can be considered as functional food but to reap the potential health benefits of that food it must be consumed on a regular basis as part of a varied diet on, at effective levels. Japan is the only country that follows a specific regulatory approval process for foods having documented health benefits. Such food are referred as Foods for Specified Health Use (FOSHU).

It is evident from an accumulating body of research that consumption of certain foods and their ingredients help in reduction of disease risk (Hasler, 1998). In recent years, foods that can alter the gut microflora is getting special interest because, it has been observed that gut microflora plays an important role in disease resistance (Khan et al., 2013). Food ingredients that affect physiological function and have positive effects on health are called biologically active components. They can be categorized in to several groups

- Polysaccharide and oligosaccharides, including functional monosaccharides, dietary fiber, small chain oligosaccharides known as prebiotics, and super sweeteners
- Polyunsaturated fatty acids, phospholipid, and choline.
- Antioxidants, Free radical scavengers, enzymatic and non-enzymatic scavengers
- Different class of vitamins, including Vitamin A, Vitamin E and Vitamin C
- Proteins and small Peptides, including glutathione, immunoglobulins, bacteriocine.
- Probiotics- Lactic acid bacteria, bifidobacterium etc
- Other active elements, such as, flavonoid, polyphenols and saponin.

Majority of the physiologically active functional food ingredients are derive from plants (Vanitha and Khan, 2019; Drzewiecki et al, 2018). However, several classes of active component of microbial origin are also used in functional food (Sanchez et al., 2012; Pham et al., 2019). These compounds shows diverse activity beneficial to human health such as antifungal, anticancer, immunosuppressive and anti-inflammatory activities. Here, we will mainly focus on microbial products in relation to their antioxidant and bio preservative properties.

Antioxidants

Free radicals play an important role in the origin of numerous diseases, including lifestyle diseases and physiological diseases like high blood pressure, cancer, diabetes, cardiovascular, neurodegenerative, etc. Free radicals act by catalysing the various toxic oxidative reactions that lead to the formation of toxic lipid peroxides. Antioxidants are biological molecules with the ability to protect vital metabolites from harmful oxidation. There are numerous medicinal plants that have been reported to possess strong antioxidant activity along with their free radical scavenging activity. Besides plants, various microbes including bacteria and fungi also possess powerful antioxidant activity. Microbial antioxidants were reported in early 1980s. To date approximately 750 species of fungi and nine genera of bacteria including both gram positive and gram negative have been identified that produce bioactive compounds having antioxidant activity (Gupta et al., 2013).

Hysashi et al, 1995, isolated an antioxidant molecule was from culture broth of *Penicillium roquefortii* IFO 5956. Structural characterization of compound characterized it as 2, 3-dihydroxy benzoic acid and its antioxidative activity was found equal to that of tert-butyl-4-hydroxyanisol (BHA). Caffeic acid (3, 4-dihydroxycinnamic acid) is another phenolic compound which is known for its pharmaceutical properties and health-promoting effects. It is generally derived from plants and synthesized by phenylpropanoid pathway. In order to improve the large scale production of caffeic acid. Lin and Yun 2012, constructed an *E.coli* strain for the production of caffeic acid through metabolic engineering. The authors found that E. *coli*.hydroxylase complex 4-hydroxyphenylacetate 3-hydroxylase (4HPA3H) was able to convert.*p*-coumaric acid to caffeic acid efficiently. Based on these finding, they modified the tyrosine pathway of for conversion of tyrosine to caffeic acid in *E.coli.* Some species of *Penicillium* and *Aspergillus* were also reported for the production of gallic acid, which is a phenolic compound and showed antioxidant properties (Sarözlü, and Vanç, 2009).

Some of the probiotic bacteria have also known to have antioxidant properties and to protect their hosts from dangerous free radicals. Hence, can be served as natural sources of antioxidants to develop unique functional foods for scavenging free radicals, thereby preventing many diseases (Gupta et al. 2013).

Nedelcheva et al. (2010) studied the antioxidant activity of *lactobacillus plantarum* strains which was used as starter culture for dried meat fermentation. It was observed that lactobacillus could preserve the colour of the meat products due to its free radical scavenging activity and provide fermented meat saugase with reduced pathogenic flora. In another study, antioxidant potential of Miang, which is a traditional fermented tea enriched with various lactobacillus sp. was studied by Klayraung and Okonogi (2009). The results suggest that Lactobacillus strains of Miang release certain compounds that are responsible for the strong antioxidant activity of the fermented tea. Antioxidant Potential of another probiotic strain *Lactococcus lactis* subsp. *cremoris* LL95 was demonstrated through scavenging of ABTS•+ radical DPPH radical scavenging activity, and ferric ion reducing antioxidant power (FRAP) assays (Ramalho et al., 2019).

Many invertebrates and Marine sponges protect themselves from environmental stress due to their excellent free radical scavenging capacity. Velho-Pereira et al, (2015), isolated and screened approximately 100 halophilic bacteria from marine sponges, corals and a single bivalve for their antioxidant activities. All the bacterial isolates belonging to the genus Chromohalobacter sp. were found excellent scavenger of superoxide free radical.

Carotenoids

Carotenoids are the pigment molecules that are responsible for imparting different colours to bacteria, plants, algae, and fungi. These are isoprenoid compounds, produced as secondary metabolites via terpenoid biosynthetic pathway. They are consisted of 35–40 carbon atoms polyene chain which is considered as backbone of the molecule. Polyene chain absorb visible light at a specific wavelength and impart colour via its absorption. Carotenoids not only provide colour to bacteria but also protect them from harmful ionizing radiation, scavenge free radicals as efficient antioxidants and also contribute in strengthening the membrane integrity. Length and structure of polyene chain also contribute in antioxidants property of molecules and it is observed that peroxylradical stabilizing ability increases with the increase in polyene chain length.

Sources of Microbial Carotenoids

Carotenoids are produced in microorganisms belonging to different groups such as psychrotrophic (*Arthrobacteragilis*), thermophilic (*Thermus thermophilus*), halophilic (*Salinibacterruber*, *Rubrobacter bracarensis*, *Halo- bacillus litoralis*), radio resistant (*Rubrobacter radiotolerans*), and other groups of bacteria and fungi also (Kumar et al., 2015). Eubacteria and yeasts belonging to the Basidiomycota are of special interest to carotenoids researchers. Carotenoid biosynthesis is a specific feature of the Rhodosporidium (Nasirian, Mirzaie, Cicek, & Levin, 2018) and Phaffia genera.

Health Benefits of Carotenoids

Consumption of carotenoid-rich foods also reported in reduced incidence of several diseases such as cancers, cardiovascular diseases, age-related macular degeneration, cataracts and diseases related to low immune function, and other degenerative diseases in different epidemiological studies (Abdel-Aal et al, 2013; Wu et al., 2015; Kyoung Sang Cho et al., 2018). Tanaka et al., 2012, showed that different classes of carotenoids like lycopene, β-carotene, β-cryptoxanthin, zeaxanthin, lutein, fucoxanthin, canthaxanthin and astaxanthin exhibit anticancer effect in both in vitro and in vivo studies. It has also been found that carotenoids having nine or more conjugated double bonds can inactivate certain reactive oxygenspecies, such as singlet oxygen. Till date, approximately 700 carotenoids have been characterized and some of them are available commercially including β-carotene, lycopene, astaxanthin, canthaxanthin, lutein and are being used in human diets. Carotenoids serve as protective pigments and function (i) to stimulate the immune system; (ii) to modulate the intracellular signaling pathways (gap junction communication) (Bertram et al, 1993); growth factors; and various types of receptors such as adhesion molecules and other physiologically processes. Some carotenoids also have role in regulation of cell differentiation, cell cycle and apoptosis (Krinsky, 2013).

Menaquinone-7, menaquinone-8, and ubiquinone-10, three isoprenoid quinines and ten carotenoids were identified from *A. bogorensis* (Antolak, et al., 2017). The main identified carotenoids in *Asaialannensis* strains were phytofluene, neurosporene, α-carotene, while for *Asaiabogorensis*, neurosporene, canthaxanthin, and zeaxanthin were noted. *Asaia* spp. were able to produce myxoxanthophyll, which has so far been identified primarily in cyanobacteria. Most interesting finding of the group carotenoids were not only accumulated by bacterial cells, but also some strains of *A. lannensis*.produced extracellular carotenoids.

Rezaeeyan et al. (2017), screened halotolerant bacterial strains isolated from tannery wastewater for carotenoid production.and identified a Gram-stain-positive coccoid, aerobic, non-endospore-forming, halotolerant bacterium strain QWT-12 that showed high production of carotenoids. 16S rRNA gene analysis classified that strain as *Kocuria.* MTT assay for extracted carotenoid was carried out against seven cancer cell lines belonging to breast, lung and prostate cancer with negative control of fibroblast cell line through six concentration levels to find out IC_{50}. Based on statistical analysis of data from MTT assay, IC_{50}.of 1, 4 and 8 mg/ml for MCF-7, (A549 and MDA-MB-468) and MDA-MB-231 respectively

Biopreservative

Bacteria are amazing producers of antimicrobial compounds since many decades, and there are enormous reports of these compounds in the scientific literature, containing novel chemical structures with diversity and modern advances in molecular genetics of bacteria (Jack and Jung, 2000). Due to increase in the multi-drug resistant bacteria, the need for novel antibiotics is essential. Humans are consuming the alternative to antibiotics like probiotic bacteria, bacteriocins (Basi-Chipalu, 2016), Lantibiotics can act as quorum sensors to trigger their biosynthesis (Dischinger et al., 2014)

The antimicrobial substances which are ribosomally synthesized are known as bacteriocins (Islam et al., 2012).They are small, heat-stable peptides showing antibacterial activity and also the producer strain is having a specific mechanism of immunity (Prieto et al., 2012). Lantibiotics are mostly produced by the gram-positive bacteria, which are ribosomally synthesized and post-translationally modified peptides (Hansen, 1993). The lantibiotics produced by gram-negative bacteria are microcins (Jack and Jung, 2000). The term lantibiotic means "lanthionine containing antibiotics" (Islam et al., 2012). Lantibiotics contain thioether bonds with amino acids like lanthionine and methyllanthionine, unsaturated amino acids like dehydroalanine and 2 amino-isobutyric acid (Alkhatib et al., 2012). Nisin was the first lantibiotic discovered in the year 1920, using as a food preservative (Basi-chipalu, 2016).

Bacillus Species Producing Compounds

Among the *Bacillus* species, *Bacillus subtilis* produces more than two dozen antibiotic(Stein, 2005).There are various enzymes released by Bacillus subtilis like proteases, amylases, endonucleases, glucose-dehydrogenase, pullulanase, and lipases (Baruzzi et al., 2011). A wide range of antimicrobial compounds are produced by the strains of Bacillus which are active against food spoilage microorganisms and also stable at high temperature, pH, resistant to enzyme treatments (Baruzzi et al., 2011). Antimicrobial compounds produced by Bacillus species are structurally unrelated to surfactin, iturin, bacteriocins, bacteriocins like inhibitory substances, and fengycins which are considered as lipopeptides (Ramachandran et al., 2014). They are an important source in the production of biochemicals, antibiotics, and insecticides, secreting heterologous proteins, nucleotides, food flavoring enhancers, amino acids like tryptophan, phenylalanine, and histidine. Vitamins like riboflavin, biotin, folic acid (Baruzzi et al., 2011). There is urgent need of antimicrobial peptides due to the emergence of drug-resistant pathogenic bacteria (Zhu et al., 2017). More than 45 antimicrobial compounds are produced by Bacillus species showing the ability of clinical importance; some compounds are active against the food spoiling pathogens and some against the plant pathogens (Baruzzi et al., 2011).

There are different classes of lantibiotics class I, which are post-translationally modified by two distinct enzymes lan B and lan C and subsequently, this pre-peptide is exported by a transporter ABC lan T. The class II and class III. Subtilin, Nisin, and Epidermin belong to class I lantibiotics which are having high antimicrobial activity against a wide range of gram-positive bacteria, whereas class III is exceptional. Lantibiotics act on the lipid II, which is the precursor of peptidoglycan cell wall (Islam et al., 2012). Lantibiotics show activity against various pathogenic bacteria including methicillin-resistant*staphylococcus aureus* and vancomycin-resistant *Enterococcus* (Islam et al., 2012) and also gram-negative pathogens having multi-drug resistance like *Acinetobacter baumanii, Enterobacteriaceae,* and *Pseudomonas aeruginosa* (Basi-Chipalu, 2016).

Table 1 Represents the Bacteriocins produced by different microorganisms

Micro-organism	Bacteriocin produced	Class	Immunity protein	Reference
Lactococcus lactis	Nisin	Class I	NisFEG, NisI	Bierbaum & sahl 2009
Staphylococcus epidermidis	Pep 5	Class I	PepI	Alkhatib et al., 2012
Bacillus subtilis ATCC6633	Subtilin	Class I	SpaFEG, SpaI	Alkhatib et al ., 2012
Bacillus subtilis A1/3	Ericin	Class I	EriFEG, EriI	Alkhatib et al., 2012
Staphylococcus epidermidis	Epidermin	Class I	EpiFEG, EpiH	Alkhatib et al., 2012
Staphylococcus gallinarium	Gallidermin	Class I	GdmFEG, GdmH	Alkhatib et al., 2012
*Staphylococcus epidermidis*K7	Epilancin	Class I	ElxI	Alkhatib et al., 2012
*Staphylococcus epidermidis*BN280	Epicidin	Class I	EciI	Alkhatib et al., 2012
Bacillus sp. Strain HIL	Mersacidin	Class I	MrsEFG	Alkhatib et al., 2012
Bacillus licheniformis	Lichenicidins 3020 Da	Class I	licFGEHI	Baruzzi et al., 2011 (Caetano et al., 2011)
Bacillus subtilis 6633	Subtilosin A 3399.7 Da	Class I	SboA	Baruzzi et al., 2011 (Zheng et al., 2000)
Bacillus subtilis 168	Sublancin 168 3877.78 Da	Class I	SunI	Baruzzi et al., 2011 (Dubois et al., 2009)
Streptococcus salivarius	Salivaricin A	Class I	Not known	Sahl and Bierbaum, 1998

Micro-organism	Bacteriocin produced	Class	Immunity protein	Reference
Streptococcus pyrogenes	SA-FF22	Class I	SA- FF22	Sahl and bierbaum, 1998; (McLaughlin et al., 1999)
Micrococcus varians	Variacin	Class I	Not known	Sahl and bierbaum, 1998
*Streptomyces*ssp	Cypemycin	Class I	Not known	Sahl and Bierbaum, 1998
Lactococcus lactis CNRZ 481	Lacticin 481	Class II	LctFEG	Alkhatib et al., 2012
Lactococcus lactis DPC3147	Lacticin 3147	Class II	LtnFEG, Ltnl	Alkhatib et al., 2012
Lactobacillus sakei L45	Lactocin S	Class II	LasJ	Alkhatib et al., 2012
Streptococcus mutans T8	Mutacin II	Class II	MutFEG	Alkhatib et al., 2012
Staphylococcus warneri ISK-1	Nukacin ISK-1	Class II	NukFEG,NukH	Alkhatib et al., 2012
Streptomyces cinnamoneus	Cinnamycin	Class II	Not known	Sahl &bierbaum 1998,
Streptomyces cinnamineus	Duramycin	Class II	Not known	Sahl &Bierbaum 1998,
Bacillus coagulans I4	Coagulin 4612 Da	Class II	pI4	Baruzzi et al., 2011, (Hyronimus et al., 1998)
Streptovermicillium ssp	Duramycin B	Class II	Not known	Sahl &Bierbaum 1998,
Streptomyces griesoluteus	Duramycin C	Class II	Not known	Sahl & bierbaum 1998,
Streptomyces ssp	Ancovenin	Class II	Not known	Sahl &Bierbaum 1998,
Actinoplanes ssp	Actagardine	Class II	Not known	Sahl & bierbaum 1998,

Many other lanthipeptides gene clusters have been identified like cinnamycin from *Streptomyces cinnamoneus*, sap B from *Streptomyces coelicolor*, sap T from *Streptomyces tendae,* Actagardine from *Actinoplanes garbadinensis*, deoxyactagardine B from *Actinoplanes liguriae*, Venezuelin from *Streptomyces venezuelae,* Microbisporicin from *Microbispora corallina*. Lanthipeptides can also be seen in bacteroidetes, chlamydiae, cyanobacteria, proteobacteria (Basi-Chipalu, 2016). The precursor peptides contain the thioether aminoacids with the enzymes involved in post-translational modifications (Dischinger et al., 2014)

Classification of Bacteriocins

According to the classification by Klaenhammer, many numbers of antimicrobial compounds are produced by Bacillus species that are grouped into Bacteriocin considered as class I which includes lantibiotic; gene encoded peptides containing lanthionine or methyllanthionine, formed through intramolecular post-translational modifications. Based on the molecular weight, biological activity and structure the lantibiotics are classified into two sub-groups A and B

Type A lantibiotics consists of 21-38 residues of amino acid exhibiting a linear structure and kills the gram-positive cells by forming pores of voltage-dependent in the cytoplasmic membrane.

Type B lantibiotics are globular, and uncharged subtilin-a pentacyclic lantibiotic consisting of 32 amino acid residues is comprised in the type A lantibiotic produced by B. *subtilis* ATCC 6633. It can withstand acidic conditions and heat treatment even at 121 °C inhibiting a wide range of bacteria that are gram-positive (Baruzzi et al., 2011).

Structure

The minimum number of amino acids required for the structural information of lantibiotics is 19 residues and to a maximum of 38. Majorly three amino acids are required for the post-translational modification of lantibiotics like the sulfhydryl amino acid cysteine, hydroxyl amino acids like serine and threonine, in rare cases the amino acids present in modified forms are lysine, asparagine, and isoleucine (Sahl and Bierbaum, 1998).

Subtilin is a pentacyclic lantibiotic containing 32-amino acid belonging to type A Lantibiotic produced by *B. subtilis* ATCC6633. The formation of pores is stabilized by lipid II interaction even at low concentrations of bacteriocins (Baruzzi et al., 2011). In stress conditions, subtilin is produced at high levels inhibiting the growth of other pathogenic bacteria. A novel subtilin-like lantibiotic is Entianin produced by Bacillus subtilis subsp. Spizizenii DSM 15029, its antibiotic and autoinduction activities are similar to subtilin (Mongkolthanaruk, 2012). The structure of subtilin is similar to nisin with exception to one specific sitei. e., protease p (lanp) is absent in the structure of subtilin operon (Alkhatib et al., 2012).

Applications

Lantibiotics have been known as the alternative to antibiotics, besides antibiotics, they can be used as food preservatives, additives, (Basi-Chipalu

2016). The antimicrobial activity of lantibiotics is promising, stable against proteases, temperature and oxidation make a new application in the health sector and related industries (Dischinger et al., 2014). Lantibiotics are frequently having a wide range of application in medicine, recently having a scope in biotechnological capabilities, showing the mode of action against multidrug-resistant bacteria by high specific activity (Cotter et al., 2005). There is an advancement in protein engineering and enzymology (Willey and Van Der Donk, 2007). They are used in the treatment of skin infections; few are having antiviral activity, amphiphilic characteristics for useful application in the drug formulation as emulsifiers and as promoters for compounds through mucosal membranes (Bower et al., 2001). It offers an advantage in the bioengineering strategies, a new niche for chemotherapeutic and as a natural food preservative (Healy and Cotter, 2017). They are having high therapeutic value (Sierra et al., 2017). From ancient, people are using various methods for preserving food, to prevent spoilage by some chemical and natural preservatives. Lantibiotics prevent the skin infections by *staphylococcus aureus* (Williams et al., 2017) lantibiotics control most of the infections caused by gram-positive bacteria having biological properties, activity against the viruses, tumor cells, having beneficial effects in the control of inflammation, blood pressure, pain relief (Gomes et al., 2017).

Conclusion

The fastest changing life style and awarness towards natural ingredients has increased the global demand functional food and neutraceuticals. However, extraction of bioactive compounds from plant, and animal source is tedious process, ans also pose ethical issues, which limits their suply. Microbe are the most divesrse groups of organims and can be used as source of large number of antioxidant compounds, vitamins, food preservatives, natural colours and dietary suplimets. Due to the fastest growing nature, ease of handing and downstream processing, microbial can be the best choice as a source of bioactive compouds for most of the industries.

References

Abdel-Aal, E.S.M., Akhtar, H., Zaheer, K. and Ali, R., 2013. Dietary sources of lutein and zeaxanthin carotenoids and their role in eye health.Nutrients, 5(4):1169-1185.

Alkhatib, Z., Abts, A., Mavaro, A., Schmitt, L. and Smits, S.H., 2012. Lantibiotics: how do producers become self-protected?.J of biotechnol, 159(3):145-154.

Antolak, H., Oracz, J., Otlewska, A., Żyżelewicz, D. and Kręgiel, D., 2017. Identification of carotenoids and isoprenoid quinones from Asaia lannensis and Asaia bogorensis. Molecules,.22 (10):1608.

Baruzzi, F., Quintieri, L., Morea, M. and Caputo, L., 2011. Antimicrobial compounds produced by Bacillus spp. and applications in food.Science against microbial pathogens: communicating current research and technological advances,.2:1102-1111.

Basi-Chipalu, S., 2016. A Review: Lantibiotics, a Promising Antimicrobial Agent.Journal of Institute of Science and Technology,.21 (1):119-128.

Bertram, J.S., 1993. Cancer prevention by carotenoids. Mechanistic studies in cultured cells. Annals of the New York Academy of Sciences,.691:177-191.

Bower, C.K., Bothwell, M.K. and McGuire, J., 2001. Lantibiotics as surface active agents for biomedical applications.Colloids and Surfaces B: Biointerfaces,.22 (4):259-265.

Brötz, H., Josten, M., Wiedemann, I., Schneider, U., Götz, F., Bierbaum, G. and Sahl, H.G., 1998. Role of lipid-bound peptidoglycan precursors in the formation of pores by nisin, epidermin and other lantibiotics.Mole microbiol, 30 (2):317-327.

Caetano, T., Krawczyk, J.M., Mösker, E., Süssmuth, R.D. and Mendo, S., 2011. Lichenicidin biosynthesis in Escherichia coli: licFGEHI immunity genes are not essential for lantibiotic production or self-protection.Appl. Environ. Microbiol. 77(14):5023-5026.

Cho, K.S., Shin, M., Kim, S. and Lee, S.B., 2018. Recent advances in studies on the therapeutic potential of dietary carotenoids in neurodegenerative diseases. Oxidative medicine and cellular longevity, 2018.

Cotter, P.D., Hill, C. and Ross, R.P., 2005. Bacterial lantibiotics: strategies to improve therapeutic potential.Current Protein and Peptide Science, 6(1):61-75.

Dischinger, J., Chipalu, S.B. and Bierbaum, G., 2014. Lantibiotics: promising candidates for future applications in health care.International Journal of Medical Microbiology, 304 (1): 51-62.

Drzewiecki, J., Martinez-Ayala, A.L., Lozano-Grande, M.A., Leontowicz, H., Leontowicz M., Jastrzebski, Z., Pasko, P., Gorinstein S. (2018). *In Vitro* Screening of Bioactive Compounds in some Gluten-Free Plants. Appl Biochem Biotechnol 186, pp847.

Dubois, J.Y.F., Kouwen, T.R., Schurich, A.K., Reis, C.R., Ensing, H.T., Trip, E.N., Zweers, J.C. and van Dijl, J.M., 2009. Immunity to the bacteriocin sublancin 168 is determined by the SunI (YolF) protein of Bacillus subtilis.Antimicrobial agents and chemotherapy,.53(2):651-661.

Earl, R. and Thomas, P.R. eds., 1994.Opportunities in the nutrition and food sciences: research challenges and the next generation of investigators. National Academies Press.

Freeland-Graves, J. and Nitzke, S., 2002. Position of the American Dietetic Association: total diet approach to communicating food and nutrition information. J of the Am Diet Assoc,.102 (1):100-108.

Gomes, K.M., Duarte, R.S. and de Freire Bastos, M.D.C., 2017. Lantibiotics produced by Actinobacteria and their potential applications (a review). Microbiology,.163 (2):109-121.

Gross, E. and Kiltz, H.H., 1973. The number and nature of α, β-unsaturated amino acids in subtilin.Biochemical and biophysical research communications, 50 (2):559-565.

Gupta, C., Prakash, D. and Gupta, S., 2013. Functional foods enhanced with microbial antioxidants.Academic Journal of Nutrition, 2(2):10-18.

Hansen, J.N., 1993. Antibiotics synthesized by posttranslational modification.Annual Reviews in Microbiology,.47 (1):535-564.

Hasler, C.M., 1998. Functional foods: their role in disease prevention and health promotion. Food Technology-Champaign Then Chicago-,52:63-147.

Hayashi, K.I., Suzuki, K., Kawaguchi, M., Nakajima, T., Suzuki, T., Numata, M. and Nakamura, T., 1995. Isolation of an antioxidant from Penicillium roquefortii IFO 5956.Bioscience, biotechnology, and biochemistry,.59 (2):319-320.

Healy, B. and Cotter, P.D., 4. Lantibiotics: Bioengineering and applications.Antimicrobial peptides: discovery, design and novel therapeutic strategies.

Hyronimus, B., Le Marrec, C. and Urdaci, M.C., 1998. Coagulin, a bacteriocin-like-inhibitory substance produced by Bacillus coagulans I.Journal of applied microbiology, 85 (1):42-50.

ILSI, 1999. Safety assessment and potential health benefits of food components based on selected scientific criteria.Isothiocyanates. Crit. Rev. Food Sci. Nutr., 39:245-257.

Islam, M.R., Nagao, J.I., Zendo, T. and Sonomoto, K., 2012. Antimicrobial mechanism of lantibiotics.

Jack, R.W. and Jung, G., 2000. Lantibiotics and microcins: polypeptides with unusual chemical diversity.Current opinion in chemical biology,.4 (3):310-317.

Kampe, B., Rösch, P. and Popp, J., 2015. Characterization of carotenoids in soil bacteria and investigation of their photodegradation by UVA radiation via resonance Raman spectroscopy.Analyst, 140(13):4584-4593.

Khan, M., Nakkeeran, E. and Umesh-Kumar, S., 2013. Potential application of pectinase in developing functional foods.Annual review of food science and technology,.4:21-34.

Klayraung, S. and Okonogi, S., 2009. Antibacterial and antioxidant activities of acid and bile resistant strains of Lactobacillus fermentum isolated from miang. Braz J of microbiol, 40(4):757-766.

Knox, J.R. and Keck, P.C., 1973. Conformation and absolute configuration of β-methyllanthionine. Biochem and biophys res commun, 53(2):567-571.

Krinsky, N.I., 1993. Micronutrients and their influence on mutagenicity and malignant transformation.Annals of the New York Academy of Sciences, 686:229-242.

Lin, Y., Jain, R. and Yan, Y., 2014. Microbial production of antioxidant food ingredients via metabolic engineering.Current opinion in biotechnology, 26:71-78.

McLaughlin, R.E., Ferretti, J.J. and Hynes, W.L., 1999. Nucleotide sequence of the streptococcin A-FF22 lantibiotic regulon: model for production of the lantibiotic SA-FF22 by strains of Streptococcus pyogenes.FEMS Microbiology Letters,.175(2):171-177.

Mercadante, A.Z., Egeland, E.S., Britton, G., Liaaen-Jensen, S. and Pfander, H., 2004. Carotenoids handbook.Britton, G., Liaaen-Jensen, S., Pfander, H., Eds.

Mongkolthanaruk, W., 2012. Classification of Bacillus beneficial substances related to plants, humans and animals.J Microbiol Biotechnol, 22 (12):1597-1604.

Nasirian, N., Mirzaie, M., Cicek, N. and Levin, D.B., 2018. Lipid and carotenoid synthesis by Rhodosporidium diobovatum, grown on glucose versus glycerol, and its biodiesel properties.Canadian Journal of Microbiology, 64(4):277-289.

Nedelcheva, P., Denkova, Z., Denev, P., Slavchev, A. and Krastanov, A., 2010. Probiotic strain Lactobacillus plantarum NBIMCC 2415 with antioxidant activity as a starter culture in the production of dried fermented meat products.Biotechnology & Biotechnological Equipment, 24(1):1624-1630.

Park, S.R., Yoon, Y.J., Pham, J.V., Yilma, M.A., Feliz, A., Majid, M.T., Maffetone, N., Walker, J.R., Kim, E., Reynolds, J.M. and Song, M.C., 2019. A Review of the Microbial Production of Bioactive Natural Products and Biologics.Frontiers in Microbiology,.10:1404.

Park, S.R., Yoon, Y.J., Pham, J.V., Yilma, M.A., Feliz, A., Majid, M.T., Maffetone, N., Walker, J.R., Kim, E., Reynolds, J.M. and Song, M.C., 2019. A Review of the Microbial Production of Bioactive Natural Products and Biologics.Frontiers in Microbiology,.10:1404.

Ramachandran, R., Chalasani, A.G., Lal, R. and Roy, U., 2014. A broad-spectrum antimicrobial activity of Bacillus subtilis RLID 12.1.The scientific world journal,.2014.

Ramalho, J.B., Soares, M.B., Spiazzi, C.C., Bicca, D.F., Soares, V.M., Pereira, J.G., da Silva, W.P., Sehn, C.P. and Cibin, F.W., 2019. In Vitro Probiotic and Antioxidant Potential of Lactococcus lactis subsp. cremoris LL95 and Its Effect in Mice Behaviour.Nutrients,.11 (4):901.

Rezaeeyan, Z., Safarpour, A., Amoozegar, M.A., Babavalian, H., Tebyanian, H. and Shakeri, F., 2017. High carotenoid production by a halotolerant bacterium, Kocuria sp. strain QWT-12 and anticancer activity of its carotenoid. EXCLI journal, 16:840.

Sahl, H.G. and Bierbaum, G., 1998. Lantibiotics: biosynthesis and biological activities of uniquely modified peptides from gram-positive bacteria.Annual Reviews in Microbiology, 52(1):41-79.

Sanchez, S., Guzman-Trampe, S., Ávalos, M., Ruiz, B., Rodríguez-Sanoja, R.O.M.I.N.A. and Jimenez-Estrada, M., 2012. Microbial natural products.Natural Products in Chemical Biology, pp.65-108.

Sar özlü, N.Y., and M. K vanç, (2009). Isolation of gallic acid-producing microorganisms and their use in the production of gallic acid from gall nuts and sumac. Afr. J. Biotechnol., 8:1110-1115

Sierra, J.M., Fuste, E., Rabanal, F., Vinuesa, T. and Vinas, M., 2017. An overview of antimicrobial peptides and the latest advances in their development.Expert Opinion on Biological Therapy,.17(6):663-676.

Stein, T., 2005. Bacillus subtilis antibiotics: structures, syntheses and specific functions. Molecular Microbiology, 56(4):845-857.

Tanaka, T., Shnimizu, M. and Moriwaki, H., 2012. Cancer chemoprevention by carotenoids. Molecules,.17(3):3202-3242

Vanitha, T. and Khan, M., 2019. Role of Pectin in Food Processing and Food Packaging. In:. Pectins-Extraction, Purification, Characterization and Applications. Eds, Martin Masuelli. Intech Open publishers.

Velho-Pereira, S., Parvatkar, P. and Furtado, I.J., 2015. Evaluation of antioxidant producing potential of halophilic bacterial bionts from marine invertebrates.Indian Journal of Pharmaceutical Sciences,.77(2):183.

Wenzel, M., Kohl, B., Münch, D., Raatschen, N., Albada, H.B., Hamoen, L., Metzler-Nolte, N., Sahl, H.G. and Bandow, J.E., 2012. Proteomic response of Bacillus subtilis to lantibiotics reflects differences in interaction with the cytoplasmic membrane.Antimicrobial Agents and Chemotherapy,. 56(11):5749-5757.

Willey, J.M. and van der Donk, W.A., 2007. Lantibiotics: peptides of diverse structure and function.Annu. Rev. Microbiol.,.61, pp.477-501.

Williams, M.R., Nakatsuji, T. and Gallo, R.L., 2017. Staphylococcus aureus: master manipulator of the skin.Cell Host & Microbe,.22 (5):579-581.

Wu, J., Cho, E., Willett, W.C., Sastry, S.M. and Schaumberg, D.A., 2015. Intakes of lutein, zeaxanthin, and other carotenoids and age-related macular degeneration during 2 decades of prospective follow-up.JAMA Ophthalmology,.133 (12):1415-1424.

Zheng, G., Hehn, R. and Zuber, P., 2000. Mutational analysis of the sbo-alb locus of Bacillus subtilis: identification of genes required for subtilosin production and immunity.Journal of Bacteriology,.182(11):3266-3273.

Zhu, M., Liu, P. and Niu, Z.W., 2017. A perspective on general direction and challenges facing antimicrobial peptides.Chinese Chemical Letters, 28(4):703-708.

15

Fermented Functional Food

[1]Akriti Dhyani, [1]Rajni Chopra, [2]Susmita Dey Sadhu and [2]Meenakshi Garg

[1]Institute of Home Economics, University of Delhi
F-4 Hauz Khas Enclave, New Delhi
[2]Bhaskaracharya College of Applied Sciences
University of Delhi, Dwarka, New Delhi

Introduction

The art of fermentation process is developed as far back as thousand years ago without understanding the role of microorganism in it. It is believed that the process of fermentation originates in Indus valley during 7000 – 8000 BC. Archives of harappan civilization showed that different clay utensils were used for the fermented product and the agriculture and animal husbandry were fully developed during that period. It is evident from the artifacts of Egypt and middle east that the practice of fermentation was common in that part (Farnworth, 2003). The earliest practice of cheese making is originated from Falcate between Tigris and the Euphrates rivers in Iraq, 8000 years back. Later, Egyptians developed the use of yeast in wine and bread making during 4000 BC (Ross et al., 2002). Further, Sumarians prepared beer from fermented barley during 1750 BC. Earlier the fermenting process was an artisan way without any knowledge of microorganism. However, the main turning point was in 1608, when Antony Van Leeuwenhoek first established the use of a microscope and revealed the existence of microorganisms. Another, major finding was Pasteurization by Louis Pasteur in 1861 AD and this was the first report that documented the role of microorganism in fermentation. Later, there was a shift observed in the production of fermented food from local communities to large scale industry. The health effect of fermented food was first reported by Roman historian Pliny, who reported that fermented milk is beneficial for gastrointestinal complains. Further, a French paediatrician proved that Bifidobacteria in food prevents infants against infection (Grill et al., 1995).

Table 1: History of some fermented food

Food	Time (BC)	Country
Cheese	6000	Iraq
Mushroom	4000	China
Wine	3500	North Africa, Europe
Bread	3000	Egypt
Beer	3000	North Africa, China
Soy sauce	3000	China, Korea, Japan
Fermented milk	3000	Middle East
Fish sauce	1000	Southeast Asia,North Africa
Pickled vegetables	1000	China, Europe
Tea	500	China

(*Source*: Campbell-Platt, 2014)

Fermentation is a metabolic process, which involve the anaerobic breakdown of carbohydrates to alcohol or organic acids by the action of microorganism and enzymes. The microorganism involved in the fermentation process are the microflora that are naturally present in substrate or must be added as a starter culture (Campbell-Platt, 1994). It is the oldest and economical method of preservation since antiquity. Along with preservation it also provides taste, forms, Flavour and enhanced the nutritional and therapeutic value.Recently, there is a huge focus on fermented food because of their therapeutic and nutritional value. One third diet of human consist of fermented food (Borresen et al., 2012). Fermented food are classified into nine broad categories on the basis of their substrate : (1) fermented cereals based (2) fermented vegetables based, (3) fermented legumes based, (4) fermented roots and tubers based, (5) fermented milk products, (6) fermented meat products, (7) fermented fish products, (8) miscellaneous fermented products, and (9) alcoholic beverages (Tamang et al., 2016).

Table 2: Common fermented food products of various country

Products	Substrate	Microorganism	Country
Yogurt	Milk,	*L. bulgaricus*, *S. thermophilus*	Greece, Turkey
Kefir	Milk, Kefir grains	*Saccharomyces cerevisiae* and *L. plantarum*	Russia
Sauerkraut	Green cabbage	*L. plantarum*	Germany
Kimchi	Cabbage	*Leuconostoc mesenteroides*	Korea
Ogi	Cereal	*Lactobacillus* sp., *Saccharomyces* sp., *Candida* sp.	South Africa
Miso	Soybeans	*Aspergillus oryzae*, Zygosaccharomyces, *Pediococcus* sp.	Japan

Products	Substrate	Microorganism	Country
Tepa	Stinkhead fermented fish	-	USA
Dosa	Fermented rice batter and lentils	*L. plantarum*	India
Fermented sausage	Meat	Lactobacillus, Pediococcus, or Micrococcus	Greece and Italy

(*Source*: Chilton et al., 2015)

Role of Fermented Food as Functional Food

In recent years, fermented food has gained immense popularity due to their nutritional and health benefits. Fermented food is also considered as Functional food. The term Functional food first proposed in Japan. Fermented food acts as functional food mainly through two mechanism, first is probiotic effect and other is biogenic effect. Probiotics are defined as the live microorganisms, administrated in certain quantities that confer health benefits to the host" (FAO/WHO 2001). Biogenic effect is indirect effect of fermented food in which they impart health promoting bioactive components (biogenics) formed during fermentation process (Stanton et al., 2005). The probiotic effect of fermented food was first observed by Metchnikoff, who reported that the consumption of fermented milks prevents the gastrointestinal problems and beneficial for gut microflora. Most widely used probiotics belongs to group Lactic acid bacteria (*Lactobacillus, Enterococcus*) and genus *Bifidobacterium.* However, various other yeasts and microbes also showed their potential as probiotics (Ouwehand et al., 2002). Various studies have proved the therapeutic effect of probiotics of fermented food such as antioxidant, antimicrobial, ant cancerous, anti-hypertensive and prevention of diabetes, osteoporosis, gastrointestinal disorder, Lactose intolerance (Tamang et al., 2009; Ping et al., 2012; Liu and Pan, 2010).

Plant Based Fermented Foods

Cereals and Legumes Based

Cereals and legumes are the most important source of nutrients in the diet of people across the world. However, they lack in certain parameters such as deficiency of some essential amino acids (lysine and methionine), low availability of starch, and presence of some antinutritional components (Phytate, tannins and polyphenol) (Bladino et al., 2003). Various method and technique are used to enhance the nutrient content of cereals and pulse such as genetic improvement, amino acid supplementation with protein concentrates or other protein-rich sources such as legumes, cooking, sprouting, milling and fermentation. Natural fermentation of cereals increases the protein content. Also, it enhanced the essential amino acids by proteolytic bacteria. It was reported that the level of lysin and crude fibre content of oats, rice, corn, wheat and millets improved by fermentation. In the same way, fermentation improves the bioavailability of B-vitamins. Several researchers have reported the reduction in antinutritional content by fermentation such as reduction of phytates in rice and during yeast fermentation of wheat (Chavan & Kadam, 1989).

Bread

Bread is the most common cereal product fermented by yeasts. The purpose of fermentation of bread is different from other fermented product, its purpose is to convert cereals grains into more consumable and palatable form. Some Archaeological data reported that bread manufacturing originated from Egypt about 3,000 year ago. The first flat breads were prepared using barley flour, which does not work well for leavened breads as it didn't give the proper airy and spongy structure. The importance of grains and bread has been reported in mythological books. Bread was also offered to ancient Greek god Artemis and Apollo during the ancient spring festival name Thargelia (Hutkins, 2006). It was reported that the folate content of bread improved 3-fold after yeast (*S. cerevisiae* CBS776)fermentation (Hjortmo et al., 2008). The cell wall of yeast (*S. cerevisiae*) is a major source of β- glucan and it was reported that β- glucan possess antioxidant activity and modulates immune function (Pelizon et al., 2005).

Idli

Idli is a common breakfast food in India especially in southern India and Sri Lanka. It is prepared from a mixture of soaked and milled rice (Oryza sativa L.) and dehulled black gram (*Phaseolus mungo* L.) in different ratios by steam cooking. Various researcher reported different proportion of black gram and rice (Nagaraju and Manohar 2000). Usually 2 parts of rice to 1 part of black gram is a common ratio in idli preparation. Black gram plays an important role in idli fermentation, it acts as a substrate and also it contains natural microflora which helps in fermentation. The lactic acid bacteria *Leuconostoc mesenteroides, Streptococcus faecalis, Lactobacillus delbrueckii, Lactobacillus fermenti, Lactobacillus lactis* and *Pediococcus cerevisiae* have been found to be responsible for the fermentation process. Acidification and leavening are two important changes that occur during idli fermentation. For leavening action microorganism involved are Heterofermentative lactic acid bacterium, *Lactobacillus mesenteroides* and for the acidification of batter, homofermentative lactic acid bacterium *Streptococcus faecalis* are responsible (Mukherjee et al.1965). Various studied have reported that idli fermentation improved the level of choline, folic acid, methionine and all the essential amino acids. Additionally, it was also reported that the fermentation of idli improved the protein efficiency ratio and reduced the antinutrients such as phytic acid, enzyme inhibitors, and flatus sugars.

Dosa

Dosa is another fermented product of south India. It is a kind of pan cake, consumed as breakfast or snacks item, in its preparation primary ingredients (black gram and rice flour) are finely ground so that its fermented batter can be spread into thin layer over hot plate.All the physical, biochemical and microbial changes of Dosa is very similar to idli.

Fermented Soy Product

Natto

Natto is traditional fermented food of Japan and it is prepared by fermenting whole yellow soybeans by *B. subtilis natto*. It is mainly consumed with boiled rice, also it is used as a seasoning agent in soup, vegetables and sea foods. Fermentation of natto helps in enhancing the amount of vitamin K. Major bioactive compounds which are responsible for the functional properties of natto are nattokinase (Fujita et al., 1993), menaquinone-7, dipicolinic acid (Murakami et al., 2003), PGA (Tanimoto et al., 2001) and levan. Natto is a richest source of menaquinone-7, it is vitamin k2 form which is mainly present in fermented foods. Various researcher has been proved a direct relation between menaquinone and bone health.

Tempeh

Tempeh is a traditional Indonesian fermented product prepared by fermenting boiled and dehulled soybeans using *Rhizopus* species. Tempeh is rich in vitamin B12. According to a study, it was reported that the soaking and fermentation process involved in tempeh preparation, reduced the anti-nutritional components such as, tempeh possess greater antioxidant ability than soybeans. In a recent study, it was reported that Tempeh exhibited greater ability to scavenge DPPH radicals as compared to soy. Also, the level of total isoflavones (Genistein and daidzein) enhanced in tempeh when compared with unfermented soybeans.

Soy Sauce

Soy sauce is a traditional seasoning originated in china, now it is widely consumed all over the world due to its unique umami taste. There are various steps involved in the preparation of Soy sauce, first step is to mix the pre-soaked soybean and roasted wheat flour and then ferment this mixture using fungi (*Aspergillus oryzae* and *A. soyae*). At this stage, it is known as Koji. Further, koji is mixed with sodium chloride at 16-18%w/v and water to make a mash called moromi. Moromi is allowed to aged and ferment using lactobacillus and yeast (*Zygosaccharomyces rouxii* and *Candida* species) at room temperature. After fermentation and aging, moromi is pressed and liquid part is further pasteurized in order to get final product. The quality of soy sauce depends on the time of fermentation, longer the fermentation time better the quality of soy sauce.

Miso

Miso is a traditional fermented soybean paste and it is used on a regular basis as a seasoning in Japanese cooking. It is a mixture of fermented soybean, rice or wheat or barley and salt. Recently, a research proved that the miso consumption suppresses the visceral fat in mice (Okouchi et al., 2019). In another study, it was found that fermented miso diet aids in the reduction of aberrant crypt foci (ACF) induced by azoxymethane (AOM) in male F344 rats, thus act as a chemoprotective agent (Ohara et al., 2002). Miso diet also helps in maintaining the blood pressure (Watanabe et al., 2006).

Table 3: Fermented Cereals and legumes products of various countries

Product	Substrate	Microorganisms	Country
Ambali	Ragi (finger millet), rice, curd	-	South India
Bouza	Wheat	-	Egypt, Turkey
Boza	Wheat, rye millet, maize, sugar	Combination of yeast and lactic acid bacteria	Bulgaria, Albania, Turkey, Romania
Chicha	Corn	*S. cerevesiae, Lactobacillus* sp., *Leuconostoc,* sp., *Acetobacter* sp., *Aspergillus* sp.	South America
Chilra	Wheat, barley,and buckwheat flour	-	South America
Dosai	Rice, black gram dhal	*Leuconostoc mesenteroids, Streptococcus faecalis, Torulopsis candida, Trichosporon pullulans*	South India

Product	Substrate	Microorganisms	Country
Dhokla	Bengal gram, rice	*L. fermentum, Leuconostoc mesenteroides, Pichia silvicola, Streptococcus faecalis, Torulopsis* sp., *Candida* sp., *Torulopsis pullulans*	North India (Gujrat)
Injera	Sorghum, tef + corn, finger millet barley	Yeasts and fungi, *Pullaria* sp., *Aspergillus* sp., *Penicillium* sp., *Rhodotorulla* sp., *Hormodendrum* sp., *Candida* sp.	Ethiopia
Kenkey	Maize	*L L. fermentum, Lactobacillus reuteri, Saccharomyces, Candida mycoderma, Penicillium, Aspergillus, Fusarium*	Ghana
Kishk	Wheat + milk	*L. plantarum, L. casei, Lactobacillus breris, Bacillus subtilis,* Yeasts	Egypt
Kisra	Sorghum, pearl millet	*Lactobacillus* sp., *S. cerevisiae, Acetobacter* sp.	Arabian Gulf, Sudan, Iraq
Mahewu	Corn meal sorghum, millet, malt, wheat flour	*L. lactis*	Africa, Arabian Gulf countries
Nan	Refined wheat flour + salt + backslop or curd	*S. cerevisiae,* Lactic acid bacteria	India, Pakistan, Afghanistan, Iran
Ogi	Maize	*L. plantarum, S. cerevisiae, C. mycoderma, Coryne bacterium, Aerobacter, Rhodotorula, Cephalosporium*	West Africa
Pozol	Maize	*L. lactis, Streptococcus suis, L. plantarum, L. casei, Lactobacillus alimentarium, Lactobacillus delbruekii*	Mexico
Rabadi	Maize flour + buttermilk	*Pediococcus acidilactis, Bacillus* spp., *Micrococcus* spp.	
Soy sauce	Soya beans and wheat	Koji fermentation: *A. oryzae* and *A. soyae* Moromi fermentation: *Zygosaccharomyces rouxii* and *Candida* sp.	Japan, China, Far East
Tarhana	Wheat flour + sheep milk + yogurt + yeast + cooked vegetables + spices	*Streptococcus thermophilus, Lactobacillus bulgaricus*	Turkey

(*Source*: Prakash, 2016)

Vegetable Based Fermented Food

Kimchi

Kimchi is a traditional fermented food of Korea, it is prepared by fermenting Chinese cabbage by Lactic acid bacteria (LAB). Other vegetables such as radish, cucumber, green onion also used for the preparation of various Kimchi. Kimchi is a low-calorie food, rich in vitamins (vitamin C, β-carotene, vitamin B complex), minerals (calcium, iron), fibres and other functional components including capsaicin, allyl compounds, gingerol, and chlorophyll.Various studies have proved that the bioactive components of kimchi such as benzyl isothiocyanate, indole compounds, thiocyanate, and β-sitosterolthat have anticancer, anti-obesity, and anti-atherosclerotic properties.

Sauerkraut

Sauerkraut is prepared by Lactic acid fermentation of shredded cabbage. It is widely consumed in Germany, china and Korea. Sauerkraut fermentation also involves yeast and moulds. It is a low-calorie food, contains high level of vitamins (vitamin A, B, C and K), minerals and phenolics compounds. Sauerkraut is very popular in USA and France; it is also known as "German Kraut" or "Cassoulet." Sauerkraut contain probiotics and antioxidants. In addition to this, it also acts as anti-carcinogenic and anti-inflammatory (Raak et al. 2014). Even Hippocrates mentioned sauerkraut as a functional and health food in his writing. In 1775, captain cook was awarded for his observation that Sauerkraut can cure scurvy (Saloheimo, 2005).

Fermented (Table) Olive

Olive fermentation is a traditional and popular method to preserve olive worldwide. According to International olive council, the consumption of table olive has increased in recent years. It is prepared by lactic acid bacteria and yeast fermentation. The role of yeast in olive fermentation is to improve the sensory attributes of olives by producing some volatile product and by reducing the phenolic compounds. On the other, lactic acid bacteria fermentation in olive showed anti-hypercholesterolemia, anti-cancer and anti-allergic effects. Fermented olives are rich in vitamins, minerals and oleic acid. Also, it contains, a natural COX-inhibitor, oleocanthal that protects against certain cancers and neurodegenerative Disorders (Parkinson and Keast 2014). (Malheiro et al. 2014) reported that fermented olives inhibit the growth of pathogenic microorganism.

Animal Based Fermented Food

Milk Based

Milk fermentation is an old practice; various fermented milk products and their health benefits are mentioned in various mythological books. The Main purpose of milk fermentation is to preserve nutrients and to extend the shelf life of milk. Also, fermenting milk with different microorganism result in variety of products with different organoleptic and functional properties. Examples of fermented milk products are curd, yoghurt, kefir and

Dahi (Curd)

Dahi (Dadhi) or curd is a very oldest fermented milk product of India. Even, it is mentioned in old text of 700 BC. It is different from western yogurt in terms of starter culture used. It can be easily prepared at household level using Lactic acid bacteria fermentation of milk of various mammals such as cow, buffalo and goat. It was reported that folic acid, riboflavin and thiamine levels increases during Dahi preparation. Protein quality of dahi increases 3-30% of milk as dahi contain all the essential amino acid (Sarkar et al. 2015). Lactose intolerant people can consume dahi, as it contains *b*-galactosidase, the enzyme helps in lactose hydrolysis (Singh and Kansal, 2003). Dahi possess probiotic effect as it contain lactic acid bacteria, protect children against diarrhoea. In addition to this, lactic acid fermentation in dahi produced some bioactive components such as diacetyl, hydrogen peroxide, and reuterin reduced the undesirable micro flora including Staphylococcus aureus, Salmonella typhi, Shigella dysenteria, Listeria monocytogens and *Escherichia coli* (Srinivasan et al., 1995).

Yoghurt

Yoghurt is very popular and widely consumed fermented milk product all over world, produced by lactic acid fermentation process using thermophilic strains of bacteria (*Lactobacillus bulgaricus* and *Streptococcus thermophilus*). Cow, buffalo and sheep milk is used for making yogurt, even in some countries mare and camel milk is also used. Nutrients composition of yogurt is similar to milk, it is rich in protein, carbohydrates, vitamins such as riboflavin (B2), thiamine (B1), cobalamin (B12), folate (B9), niacin (B3) and vitamin A and minerals such as phosphorous, calcium, zinc and magnesium. However, the biological value of protein is high, as it contains all the essential amino acids. Some researchers also proved that the bioavailability of calcium also enhanced in yoghurt after fermentation. Along with the high nutritional composition, yoghurt provide various health benefits to human health. Yogurt play an important role in maintaining the optimum intestinal microflora, which is associated with gut health. Yogurt act as a Probiotics carrier; Mostly all the probiotics food products available in market contain Lactobacillus and *Bifidobacterium species.* Several human and animal studies have proved the therapeutic effect of probiotics on health such as hypercholesterolemia, anticarcinogenic activity; protection against urogenital infections, constipation, diarrhoea, colon cancer and osteoporosis. Also, yogurt helps in maintaining normal intestinal flora and immune system. Numerous studies have proved that yoghurt is good for a person suffering from lactose intoleranceas during fermentation lactose hydrolysis into glucose and galactose by the lactic acid bacteria. In addition, low calcium intake is directly associated with lactose intolerance, hence high-level calcium of yogurt prevents against lactose intolerance.

Kefir

Kefir is a fermented milk beverage originated from Caucasian mountains and Tibet,its name derived from Turkish word 'Keyif' which means 'good feeling'. Kefir is made by fermentation of milk with kefir grains. Kefir grain contains several species of both bacteria and yeast, coexist as symbiotic association, in a closed matrix of polysaccharide and protein. This property makes it unique from other fermented milk product. Recently, the consumption of kefir has increased due to its distinct flavour and various health effects. The end products of kefir fermentation are lactic acid, ethanol and CO2, along with some minor components including diacetyl, acetaldehyde, ethyl and amino acids. Kefir grains naturally contain enzymes β-galactosidase, which insufficiency in the intestine leads to lactose intolerance, hence aids in lactose intolerance.

Fermented Fish Products

There are various methods to preserve the fish including drying, salting, smoking and fermentation. Fish fermentation as a mean of preservation, have been practiced from ancient time. (Beddows, 1997) reported the consumption of fermented fish during the Yayoi period (300 BC to 300 AD) in Japan. Fermented fish products are mainly consumed in Southeast Asian countries. The practice of fermenting fish is basically done to improve the perishability of fish. Naturally occurring enzymes and bacteria are the responsible for fish fermentation.Now a days, various fermented fish products are available all over the world.

Classification of fermented fish product based on the nature of final product such as

- When fish retain their original shape examples are shidhal (Bangladesh)
- Fish reduced in the form of paste examples are Nga-pi (Bangladesh), fish miso (Japan)
- Fish is reduced in liquid form examples are fish sauce Nouc-mam (Vietnam), Nampla (Thailand) shott-suru (Japan) (Giri et al., 2010). Example of various fermented fish products is given in table 4.

Table 4: Fermented Fish Products of Different Country

Product	Substrate	Microorganism	Country
Garum (sauce)	Tunnus thynnus, Scomber scombrus	-	Roman
Hakarl (Whole slices)	Greenland Shark (Somniosus microcephalus), salt	Moraxella, *Acinetobacter, Lactobacillus* sp.	Iceland
Surstromming (Whole slices)	Baltic herring, salt	Halanaerobium	Sweden
Rakfisk (Whole slices)	Salmonid freshwater fish, salt	Lactobacilli	Norway
Momoni (Whole slices)	Snakehead fish, salt, palm sirup, roasted rice	*Micrococcus, Staphylococcus aureus, Staphylococcus* sp., *Bacillus sp., Lactobacillus* sp., *Pseudomonas, Pediococcus, Klebsiella, Debaryomyces, Hansenula, Aspergillus.*	Ghana
Nam-pla (Sauce)	*Stolephorus spp. Ristrelliger* spp. *Cirrhinus* spp., salt	Homofermentative tetrad forming lactic acid bacteria	Thailand
Budu (Sauce)	Raw anchovies (*Stolephorus* spp.) and salt	*Micrococcus luteus, Staphylococcus arlettae*	Malaysia
Bakasang (Sauce)	Sardines (*Engraulis japonicus*), salt	*Micrococcus, Streptococcus, Pediococcus* sp.	Indonesia
Yu-lu (Sauce)	Anchovies (*Engraulis japonicus*), snakehead fish (*Channa asiatica*)	Halotolerant, LAB and Yeast	China
Shidal	Whole Punti fish (*Puntius sophore*), phasa fish (*Setipinna phasa*)	*Staphylococcus; Micrococcus, Bacillus*	India
Ngari (Whole slices)	*Puntius sophore*	*Lactococcus lactis* subsp. *cremoris, L. plantarum, Bacillus subtilis, Candida, Saccharomycopsis*	India
Jeotgal (Whole slices)	Shrimp, shellfish, fish, salt	*Achromobacter, Bacillus, Brevibacterium, Flavobacterium, Halobacterium, Leuconostoc, Micrococcus, Pseudomonas, Staphylococcus, Sarcina, Saccharomyces, Torulopsis*	Korean

Product	Substrate	Microorganism	Country
Katsuobushi (Whole slices)	Skipjack (*Euthynnus pelamis* or *Katsuwonus pelamis*), eastern little tuna (*E. affinis*), frigate mackerel (*Auxis rochei*), frigate tuna (*A. thazard*), oriental bonito (*Sarda orientalis*)	*Aspergillus, Eurotium*	Japan
Narezushi (Whole slices)	Crucian carp, horse mackerel, chub mackerel, salt, cooked rice	*Lactobacillus acidipiscis, Lactobacillus versmoldensis*	Japan

(*Source*: Zang et al., 2019)

Various study has proved the antioxidant activity of fermented fish. According to a study, Shidals salt free fermented fish of northeast India, could be used as a potential source of nutrients and natural antioxidants (Majumdar et al., 2016). Two novel peptides were identified as Ala-Ile-Pro-Pro-His-Pro-Tyr-Pro (AIPPHPYP) and Ile-Ala-Glu-Val-Phe-Leu-Ile-Tre-Asp-Pro-Lys (IAEVFLITDPK) from fermented fish Loma and these peptides possess strong antioxidant activity (Nazafian and babji, 2016). Also, katsuobushi oligopeptides derived from katsuobushi showed anti- hypertensive effect when given at a dose of 3g/ day to borderline and mildly hypertensive subjects (Fujita et al., 2001). Jeotgals, traditional salted fermented fish of Korea protect against liver hepatocellular carcinoma cell HepG2 and can be considered as potential anti-carcinogens (Koo et al. 2016).

Fermented Meat Products

Meat and meat products are the important part of human diet in all communities in the world. Meat products are rich in protein, essentials amino acids, vitamins (especially the B group vitamins) and minerals such as zinc and iron with high bio availability. meat fermentation is a preferred method meat safe for consumers in terms of stability, transportation and storage and mostly carried out starter culture, starter culture is a microbiological culture of defined organisms consisting of cultivation medium or nutrient liquids colonized by living microorganisms used for fermentation. The starter cultures are classified into two groups, Lactic acid Bacteria, responsible for acidification process (*Lactobacillus, Pediococcus*), and flavoring microorganisms, capable of reducing nitrite (*Staphylococcus, Kocuria, Debaryomyces, Penicillium*). Additionally, *Debaryomyces hansenii* is the dominant yeast species generally present in naturally fermented sausages and is available as a starter culture (Jessen 1995). Singh et al. (2012) reported various benefits of fermented meat products over traditional meat products:

- Improved sensory attributes such as flavour, taste, aroma, and colour.
- Fermented meat products are tender in nature due to action of proteolytic enzymes produced from starter cultures at the time of fermentation.
- The storage life of fermented meat products is longer due to acidification, low water activity, and production of bacteriocins.
- Fermented meat products are safe for human consumption due to preventing growth of pathogenic and spoilage microorganisms.
- Fermented meat products are easier to prepare, and are energy-efficient.

Sausages

Sausages preparation is very traditional practice originated in the Mediterranean region. Sausage word is derived from a Latin word Salsus means salted. Earlier when salting method was used to preserve meat and meat product. Prehistoric societies made sausages some thousands of years ago; for example, sausages were used by Egyptians as early as 2000 BC. "circelli," "tomacinae," "butuli," are the sausage prepared by the Romans and during that period sausages were consumed by the poor population of Romans. There are types of sausages semi dry and der fermented sausages. The main difference between Semi dry and dry sausages is their fermentation time. Preparation of dry sausages involved chopping, mincing meat, adding seasoning, mixing, ripening, curing and sometimes smoking. It doesn't involve cooking. Sucuk, dry-fermented spicy sausage, traditional meat product of Turkey prepared from sheep/goat/cattle meat. The semi-dry sausages are prepared from ground pork or beef or a combination of beef and pork both. The long shelf life of semi-dry sausages is attributed to heat treatment, high salt and moisture ratio, and low pH. Examples of semi dry sausages are Thuringer, summer sausage, and cervelat (Kumar et al., 2017).

Fermented Meat Products as Probiotic Carriers

Dry-fermented sausages are prepared without involving heat thus showed suitable carrier for beneficial live bacterial culture (Probiotics). However, sausage matrix and fat also provide a protective environment to probiotic cells during processing (Klinberg et al., 2005). Salami, a dry fermented sausage of Germany, contains three intestinal live LAB strains, viz. Lactobacillus acidophilus, *Lactobacillus casei,* and *Bifidobacterium* spp. The effect of fermented probiotic sausage CD4, T helper cells, bioactive compounds, conjugated linoleic acid (CLA), antibodies against oxidized low-density lipoprotein (LDL), and phagocytosis index (Bunte et al., 2000). Rebucci et al (2007) reported that daily consumption of 50 g of raw cured probiotic sausage effectively modulates the immune system.

References

Beddows, C. G. 1998. Fermented fish and fish products. In: Microbiology of fermented foods, 416–440. Boston, MA: Springer.

Blandino, A., Al-Aseeria, M.E., Pandiellaa, S.S., Canterob, D., Webba, C., Cereal-based fermented foods and beverages. Food Research International, 36, 527–543, 2003

Borresen, E.C., Henderson, A.J., Kumar, A., Weir, T.L., Ryan, E.P., 2012. Fermented foods: patented approaches and formulations for nutritional supplementation and health promotion. Recent Patents on Food, Nutrition & Agriculture 4 (2), 134–140.

Campbell-Platt, G., Fermented food, Encyclopedia of food microbiology, 1, 834-838, 2014

Campbell-Platt, G., Fermented foods: a world perspective. Food Research International, 27, 253, 1994.

Chavan, J. K., & Kadam, S. S., Critical reviews in food science and nutrition. Food Science, 28, 348–400, 1989.

Chilton, S. N., Burton, J. P., Reid, G., Inclusion of Fermented Foods in Food Guides around the World, Nutrients, 7, 390-404, 2015.

Farnworth, E.R. Handbook of Fermented Functional Food, C.R.C press, New York, 2005.

Fujita, H., T. Yamagami, and K. Ohshima. 2001. Effects of an ace inhibitory agent, Katsuobushi oligopeptide, in the spontaneously hypertensive rat and in borderline and mildly hypertensive subjects. Nutrition Research, 21 (8), 1149–58, 2001

Fujita, M., Hong, K.,.Ito, Y., Fujii, R., Kariya, K., Nishimuro, S.,Thrombolytic effect of nattokinase on a chemically induced thrombosis model in rat. Biol. Pharm. Bull,.18, 1387-1391, 1995.

Giri, A., Osako, K., Okamoto, A., Ohshima, T., Olfactometric characterization of aroma active compounds in fermented fish paste in comparison with fish sauce, fermented soy paste and sauce products. Food Research International 43 (4):1027–40, 2010.

Grill, J.P., Manginot-Durr, C., Schneider, F., Ballongueet J., Bifidobacteria and probiotic effects: action of Bifidobacterium species on conjugated bile salts. Curr Microbiol,.31, 23-27, 1995.

Hjortmo, S., Patring, J.,Jastrebova, J., Andlid, T., Biofortification of folates in white wheat bread by selection of yeast strain and process, International Journal of Food Microbiology, 127, 32-36, 2008

Hutkins, R. W., Microbiology and Technology of Fermented Foods, Blackwell Publishing, 261-298, 2006.

Koo, O. K., S. J. Lee, K. R. Chung, D. J. Jang, H. J. Yang, and D. Y. Kwon. Korean traditional fermented fish products: Jeotgal. Journal of Ethnic Foods 3 (2), 107–16, 2016.

Kumar, P., Chatli, M. K., Verma, A. K., Mehta, N., Malav, O. P., Kumar, P., Sharma, N., Quality, functionality, and shelf life of fermented meat and meat products: A review, Critical Reviews in Food Science and Nutrition, 57(13): 2844-2856, 2017.

Liu, C.F., and Pan, T. M. (2010). In vitro effects of lactic acid bacteria on cancer cell viability and antioxidant activity. J. Food Drug Anal. 18, 77–86.

Majumdar, R. K., D. Roy, S. Bejjanki, and N. Bhaskar. 2016. Chemical and microbial properties of shidal, a traditional fermented fish of northeast India. Journal of Food Science and Technology 53(1):401–10, 2016.

Malheiro, R., P. Mendes, F. Fernandes, N. Rodrigues, A. Bento, and J. A.Pereira. 2014. Bioactivity and phenolic composition from natural fermentable olives. Food & Function 5 (12), 3132–42.

Mukherjee, S. K., Albury, M. N., Pederson, C. S., Vanveen, A. G. and Steinkraus, K. H., Role of Leuconostoc rnesenteroides in Leavening the Batter Of Idli, A Fermented Food of India. Appl. Microbiol. 13, 227-231, 1965.

Murakami, K., Tanemura, Y., Yoshino, M., Dipicolinic acid prevents the copper-dependent oxidation of low-density lipoprotein, J. Nutr. Biochem,.14: 99-103, 2003.

Nagaraju, B. D. & Manohar, B., Rheology and particle size changes during Idli fermentation, Journal of Food Engineering, 43:167-171, 2000.

Ohara, M.; Lu, H.; Shiraki, K.; Ishimura, Y., Uesaka, T., Katoh, O., Watanabe, H. Prevention by long-term fermented miso of induction of colonic aberrant crypt foci by azoxymethane in F344 rats. Oncol. Rep., 9: 69–73, 2002.

Okouchi, R., Sakanoi, Y., Tsuduki, T., Miso (Fermented Soybean Paste) Suppresses Visceral Fat Accumulation in Mice, Especially in Combination with Exercise. Nutrients, 11, 560, 2019.

Ouwehand, A.C., Salminen,S., Isolauri, E. (2002). Probiotics: an overview of beneficial effects. Antonie VanLeeuwen, 82: 279–289.

Parkinson, L., and R. Keast. 2014. Oleocanthal, a phenolic derived from virgin olive oil: a review of the beneficial effects on inflammatory disease. International Journal of Molecular Sciences, 15 (7):12323–12334

Pelizon, A.C., Kaneno, R., Soares, A.M.V., Meira, D.A., Sartori, A., Immunomodulatory activities associated with b-Glucan derived from.Saccharomyces cerevisiae. Physiological Research, 54, 557-564, 2005.

Ping, S. P., Shih, S. C., Rong, C. T., and King, W. Q. (2012). Effect of isoflavone aglycone content and antioxidation activity in natto by various cultures of Bacillus subtilis during the fermentation period. J. Nutri. Food Sci. 2:153.

Prakash, J., (2015). Regulating Safety of Traditional and Ethnic Foods.

Raak, C., Ostermann, T., Boehm, K., Molsberger, F., Regular Consumption of Sauerkraut and Its Effect on Human Health: A Bibliometric Analysis, Global Adv Health Med. 23(6), 12-18, 2014

Ross, R.P., Morgan, S., Hill, C., Preservation and fermentation: past, present and future. International Journal of Food Microbiology, 79, 3 – 16, 2002.

Saloheimo, P.,.Captain Cook used sauerkraut to prevent scurvy.Duodecim Laaketieteellinen Aikakauskirja. 2005; 121(9): 1014–5.

Sarkar, P., Kumar, L., Dhumal, C., Panigrahi, S.S., Choudhary, R., Traditional and ayurvedic foods of Indian origin, J Ethn Foods, 2, 97-109, 2015.

Singh, R., Kansal, V.K., Augmentation of immune response in mice fed with dahi fermented milk containing Leuconostoc citrovorum and Lactococcus lactis. Milchwiss. 58, 280–282, 2003.

Srinivasan, R., Sarkar, S., Pramanik, K.K., Kuila, R.K., Misra, A.K., Isolationand characterization of Lactococcus species producing bacteriocins. Indian J. Dairy Sci. 48: 596–602, 1995.

Stanton, C., Ross R.P., Fitzgeral, G.F., Sinderen, D.V., Fermented functional foods based on probiotics and their biogenic metabolites, Current Opinion in Biotechnology, 16,198–203, 2005.

Tamang, J. P., Watanabe, K., Holzapfel, W. H., Review: Diversity of Microorganisms in Global Fermented Foods and Beverages, Frontiers in Microbiology. 7: 377, 2016.

Tamang,J.P., and Fleet, G.H. (2009). "Yeasts diversity in fermented foods and beverages," in Yeasts Biotechnology: Diversity and Applications, eds T. Satyanaray Anaand G. Kunze (New York: Springer), 169–198.

Tanimoto, H., Mori, M., Motoki, M., Torii, K., Kadowaki, M., Noguchi, T., Natto mucilage containing poly-γ-glutamic acid increases soluble calcium in the rat small intestine, Biosci. Biotechnol. Biochem 65, 516-521, 2001.

Watanabe, H., Kashimoto, N., Kajimura, J., Kamiya, K. A miso (Japanese soybean paste) diet conferred greater protection against hypertension than a sodium chloride diet in Dahl salt-sensitive rats. Hypertens. Res, 29:731–738, 2006.

Zang, J., Xu, Y., Xia, W., Regenstein, J.W., Quality, functionality, and microbiology of fermented fish: a review, Critical Reviews in Food Science and Nutrition, 60 (11): 1-15, 2019.

16

Utilization of Dairy Industry Waste for Production of Value Added Functional Food Products to Combat Covid-19 Pandemic and Other Chronic Diseases

***Saloni*[1], *Dinesh Chandra Rai*[1], *Veena Paul*[1], *Rani Kumra*[2] and *Abhishek Dutt Tripathi*[1]**

[1]Department of Dairy Science and Food Technology
Institute of Agricultural Sciences, Banaras Hindu University
Varanasi, Uttar Pradesh
[2]Department of Home Science, Mahila Mahavidyalaya
Banras Hindu University, Varanasi, Uttar Pradesh

Introduction

COVID-19, a worldwide pandemic, addresses a deflationary stun for the worldwide economy. The outbreak of coronavirus has had a significant impact on public health worldwide and all stages of various industries' supply chain and value chain. The 2019 novel coronavirus (2019-nCoV) or the severe acute respiratory syndrome coronavirus 2 (SARS-CoV-2) is rapidly spreading from its origin in Wuhan City of Hubei Province of China to the rest of the world. The infection began in bats and was communicated to people by inhalation or contact with infected droplets, and its infection time frame goes from 2 to 14 days. Primary manifestations in hospitalized patients with COVID-19 include fever (98%), cough/hack (76%), and myalgia or weakness (44%). Severe complications include acute respiratory distress syndrome (ARDS), RNAemia, acute cardiac injury, and multiple organ failure. However, exaggerated immune responses may lead to Cytokine Storm. Cytokine storm syndrome includes ARDS, shock, and multi-organ failure (Zheng et al., 2020). Many people are asymptomatic also, and the severity of this disease varies from person to person. The fatality rate is estimated to range from 2 to 3%. Typical/low white cell checks can analyze it with raised C-responsive protein (CRP), periodic modernized tomographic chest examination. The infection spreads quicker than its two predecessors, the SARS-CoV and Middle East respiratory disorder Covid (MERS-CoV); however, it has lower casualty. The worldwide effect of this new scourge is yet questionable (Singhal, T. (2020).

Antiviral medications for overseeing contaminations with human coronaviruses are not yet endorsed. Remdesivir (RDV) is an investigational compound with an expansive range of antiviral activities against RNA infections, including severe acute respiratory syndrome-CoV and the Middle East respiratory syndrome (MERS-CoV). RDV is a nucleotide analog inhibitor of RNA-dependent RNA polymerases (RdRps). The selectivity esteem for RDV-TP is more productive than ATP and two other nucleotide analogs. When incorporated at a position I, the inhibitor caused RNA blend capture at position i+ 3. Subsequently, the possible activity system is postponed RNA chain termination (Gordon et al., 2020).

Another approach is to boost immunity, and this can be achieved by the intake of nutraceuticals and functional food in the diet. Optimal nutrition can improve well-being and alleviate the danger and dreariness of coronavirus infection 2019 (COVID-19). Nonetheless, supplementation with nutrients C and D, just as zinc and selenium, was featured as possibly beneficial for people with, or in danger of, respiratory viral contaminations or for those who have a nutrient deficiency is detected (de Faria et al., 2020). By using plenty of water, minerals like magnesium and zinc, micronutrients, herbs, food rich in vitamins C, D, and E, and a better life style, one can promote health and overcome this infection (Arshad et al., 2020). The dairy industry represents a broad scope for healthy and nutritious product formulation.

Nowadays, dairy wastage and its gathering turn into a fundamental issue worldwide because of the consistent increment of the total population. The exponential development in dairy waste is forcing genuine dangers to our general public like natural contamination, well-being danger, and dumping land shortage. Consistently, around the world, dairy waste of about 4-11 million tonnes is delivered in the climate, being a risk for biodiversity (Ahmad et al., 2019). At present, different sorts of approaches are explored in dairy waste handling and the executives for cultural advantages and applications.

World whey yield is around 180 million tons containing some 1.5 million tons of high worth protein and 8.6 million tons of lactose. Every year, there is a 1-2 percent increment in total whey production, but less than 50 percent of total whey produced is used or processed with different technologies and processes. About 40 percent of whey is disposed of around the world, leading to substantial valuable nutrients loss. It shows that a significant amount of milk nutrients is not ultimately used for humankind and such inefficient use leads to profound sustainability implications (Panghal and Patidar, 2018).

Dairy waste management has one of the major concerns because the dairy sector has economic importance in various industries of agricultural sectors. India is the world's biggest milk producer, with 22 percent of worldwide production, trailed by the United States of America, China, Pakistan, and Brazil. (http://www.fao.org/dairy-production).

In terms of milk production, India ranked first and grew at 6.5 percent to 187.7 million tonnes in 2018-19, but only a small part of produced milk is processed. The dairy industry is expected to proliferate and have the waste generation, and related environmental problems are also assumed increased importance by considering the increased milk and its by-product demand to produce healthy and immune booster product, the dairy industry is expected to grow rapidly and have the waste generation and related environmental problems are also assumed increased importance.

Sources and Composition of Dairy Waste

Dairy industry wastes are originated from different dairy processing units like cheese plant, casein plant, ice cream plant, condensed milk, and dried milk plant, etc. due to spills and leakage of product or by-product, residual milk or milk products, wash solutions from equipment, and machines, etc. The primary wastes are whey, sludge, and wastewater by processing, cleaning, and sanitization (Ahmad et al., 2019).

These wastes are composed of solid that includes carbohydrate, fat, protein, minerals and rest part of the water that may include sweeteners, salt, sanitizers, cleaners, gum, flavoring, etc. some minor constituent also includes nitrogenous compounds, phosphorus, and chlorides, etc. The phosphorus substance of milk is around 1000 mg/L, while in whey, it is around 450-575 mg/L.

Milk contains nutritive mixes as well as biologically active ones, like caseins and whey proteins. Caseins constitute 80% of the total protein present in milk; the other 20% is comprised of the whey proteins, including α-lactalbumin, β-lactoglobulin, and immunoglobulins (De Jesus et al., 2015).

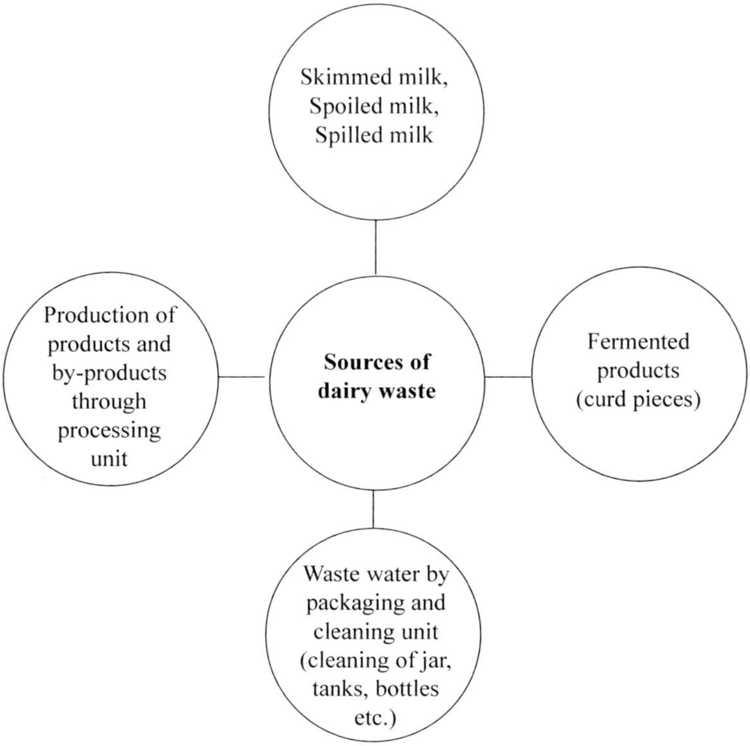

Fig. 1. Different sources of dairy waste from dairy processing waste

Whey is produced as a by-product during cheese and casein manufacture containing lactose and protein, and it could be separated by ultrafiltration, nanofiltration, diafiltration, and different advanced separation techniques (Das et al., 2016). Dairy wastewater has fats, lactose, whey proteins, and supplements, which assume a significant part in increasing the biological oxygen demand (BOD) of water. Wastewater also contains cleansers and sanitizing agents, which are consequences of the cleaning cycle and increase chemical oxygen demand (COD). There are many ways to reduce the BOD and COD. Both aerobic and anaerobic technologies have been used. In contrast, membrane technology is alternatives to biotreatment being used for BOD reduction in dairy wastewaters, which replaces secondary clarifiers in the waste treatment plants with membranes.

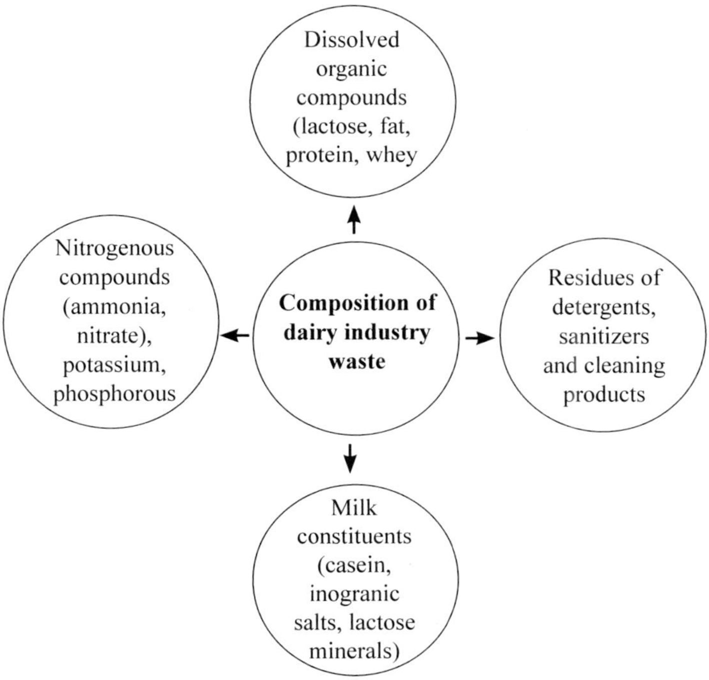

Fig. 2: Composition of dairy industry waste

Dairy wastewater is distinguished by high BOD (40–48,000 mg/L) and COD (80-95,000mg/L) and pH varying from 4.7 to 11. It is white with an unsavory scent and turbid character (whey has a yellowish-green tone). Due to the fine curd found in cheese waste, the suspended matter present in milk waste is quite prominent.

Treatment of Indian Dairy Wastes

Milk processing consumes a large amount of water and generates 6–10 liters of effluent per liter of milk processed. An effluent volume is approximately four times the volume of processed milk (Gopinathan & Thirumurthy 2012). The main objectives of dairy waste management are reducing or removing organic content, minerals, and pathogenic microbes or parasites. Pretreatments include chemical pretreatments, heat pretreatments, ultrasound pretreatments, turbulence use, etc. (Saprsh and Parag 2017).

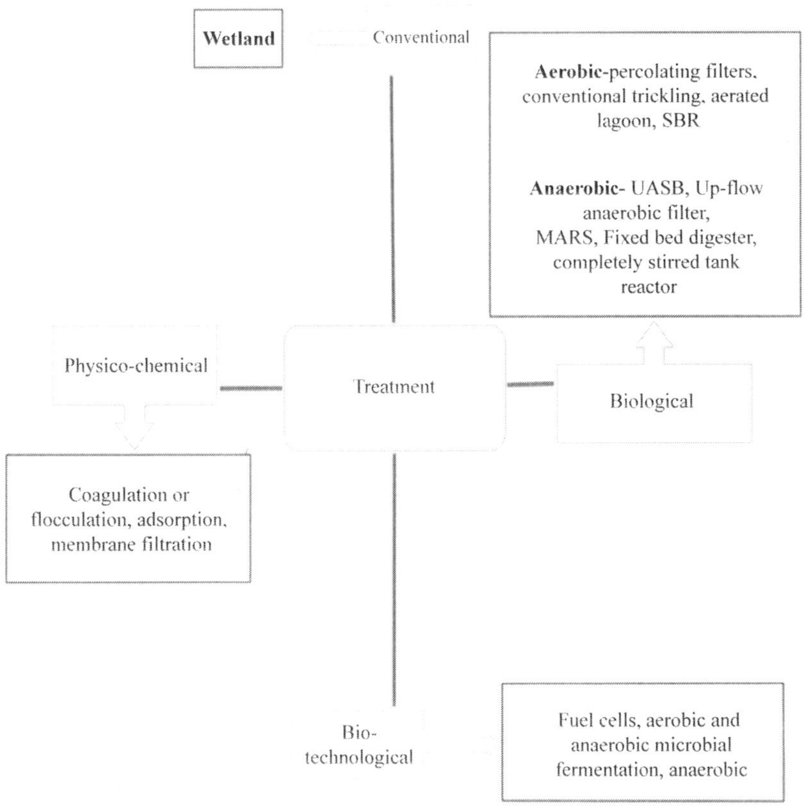

Fig. 3: Different approaches for treatment of dairy waste

Conventional Treatment

The conventional biological method comprised of the construction of centralized facilities. These establishments had high initial speculation cost, a significant expense for the continuous and proper activity, the significant expense of maintenance, were energy devouring, brought about high carbon dioxide outflows, delivered smells and noise (Ahmad et al., 2019).

Wetland Treatment

Wetlands are considered sustainable wastewater treatment, as they have the same function as the conventional treatments (effective wastewater treatment) but in a more economical, environment friendly, and energy-efficient form. This treatment utilized wetland plants (hydrophytes), aggregate materials and microorganisms, and is particularly ideal for usage with dairy effluents due to simple construction and absence of sludge reusing (Ahmad et al., 2019).

Physico-Chemical Process

Electrocoagulation (EC)

Electrocoagulation (EC): Electrocoagulation with Aluminum anode is an advantageous route for the treatment of dairy waste, yet COD's expulsion efficiency through this strategy is just 61%. Contrasted with synthetic coagulation, the effluents treated by electrocoagulation have a lower conductivity, which can empower the mechanical reusing of water. Due to the high expulsion paces of phosphorus and nitrogen by electrocoagulation, this electrolytic method can be seen as a reciprocal cycle to methanisation (Tchamango et al., 2010).

Electrocoagulation (EC) depends upon the electrochemical disintegration of sacrificial metal (typically iron or aluminum) induced by the passage of electric flow into a solvent or insoluble species that improve the coagulation, the adsorption, or the precipitation of dissolvable or colloidal contaminations in different water and wastewater effluents in the batch mode yet additionally in the continuous mode. The process helps in the significant removal of suspended colloidal particles. A similar study on the EC process for the treatment of dairy wastewater was performed by Chezeau et al., 2020. They found that final removal yield in the range of 55% COD, 60% total organic carbon, 90% total nitrogen, and nearly 100% turbidity during the analysis of waste.

In this kind of treatment, low-cost food grade sodium lignosulphonate (Na-lignosulphonate) can be utilized to recuperate proteins and lipids from dairy wastewater and reduce the BOD. The colloidal particles of the wastewater sample were precipitated by changing Na-lignosulphonate concentration (0.002–0.032%, w/v) and pH of the medium (1.0–3.5). Experiments were performed at ambient (22 °C) and mildly elevated (40 °C) temperatures. At the optimum concentration (0.016%, w/v) of Na-lignosulphonate, the highest turbidity removal (98%) was accomplished at pH 3.5 and both temperatures. This technique recovered 96% lipids and 46% proteins (mostly caseins) and reduced the BOD by 73% at 22 °C. Na-lignosulphonate is found to be a powerful coagulant attributable owing to its capacity to be readily involved with positively charged colloidal particles at acidic pH (Kurup et al., 2019).

Dairy Wastewater Treatment Using Water Treatment Sludge as Coagulant

Water treatment sludge was used as a coagulant to treat synthetic dairy wastewater in the pH range of 4–10. Turbidity, COD, BOD, TSS, and TDS expulsions from the synthetic dairy wastewater were discovered to be around 93, 65, 67, 84, and 85%, respectively, at the ideal conditions. Water treatment sludge was found to perform even better than other traditional coagulants utilized for the same. Results demonstrated that it could substitute the common coagulants wholly or partially in the essential treatment of dairy wastewater. The utilization of water treatment sludge at dairy wastewater treatment plants would give sustainable sludge management and cost-effective dairy wastewater treatment (Suman et al., 2018).

Electrochemical and Electroflotation Processes

The treatment of simulated dairy wastewater (SDW) was performed by electrochemical (EC) strategy by utilizing an iron electrode. COD, TS, TN, and turbidity removal efficiencies at optimum conditions were found to be 70%, 48.2%, 92.75%, and 99.8%, respectively. The cost of both electricity and the electrode consumption required to run the system at optimum conditions reduces the COD of 1 m3 of dairy wastewater from 0.43–15.2 kg/m3 to 0.129–4.6 kg/m3 is in the range of 0.072–2.55 $. The removal of TGA/DTA investigation indicated that created sludge and rubbish can be dried and utilized as a fuel in the boilers/incinerators or can be utilized to create fuel-briquettes. (Kushwaha et al., 2010).

The electroflotation innovation is compelling in eliminating colloidal particles, oil, grease, and organic pollutants from wastewater. This procedure can be utilized as a pretreatment method of milk wastewater and its contribution to reducing pollutants in the milk processing industry. Furthermore, the process of electroflotation and electrochemical methods seems to be effective in remediating varieties of pollutants of different chemical compositions and can achieve high treatment efficiency (Mohammed, 2017).

Adsorption

Among the various physicochemical treatment methods, adsorption is attractive for the removal of organic compounds in wastewaters. Chitosan can be successfully used as an adsorbent for the treatment of dairy wastewater. Its non-toxicity, biodegradability, and environmentally friendly quality make it an effective bio-material in the expulsion of pollutants from dairy effluent. Batch study indicates that chitosan can remove 79 % COD. However, no significant effect was noticed in the percentage removal of COD after a dosage of 150 mg/L. Experimental results show that COD reduction increased with increased dosage of adsorbent up to 150 mg/L, at a contact time of 50 min and a stirring speed of 50 rpm. The parameter reductions were best effective at an optimum pH of 5 (Devi et al., 2012).

Treatment of synthetic dairy wastewater can likewise be performed using adsorption onto a multi-walled carbon nanotube. The effects of contact time, initial concentration, temperature, adsorbent dosage, and pH on the removal of dairy wastewater were also investigated. The equilibrium data were analyzed by the Langmuir, Freundlich, Temkin, Handersen, and Harkins-jura, and Halsey models, which explained that the Langmuir model is more suitable to describe the dairy wastewater adsorption than other models (Moradi and Maleki, 2013)

For wastewaters' treatment, cost-effective adsorbents such as fly ash, bagasse, wheat straw dust, sawdust, and coconut coir were used to reduce the TDS (total dissolved solids) from the dairy industry effluent wastewater. PAC (powdered activated carbon) was also used, and the results were compared. Sorption data have been correlated with both the Langmuir and the Freundlich adsorption isotherm models. The Freundlich static isotherm model is found applicable to all the six adsorbents for removing TDS from the dairy wastewater (Rao et al., 2002).

Membrane Treatment

The typical membrane separation processes are microfiltration, nanofiltration, ultrafiltration, reverse osmosis, and electro-dialysis. High feasible product recovery is possible with these methods. Nano Filtration (NF) membrane treatment is a suitable option compared to the conventional treatment by Reverse Osmosis (RO) because it can work at lower pressures, lower-energy consumption, and higher permeate recoveries than RO. The study detailed the recovery of lactose and milk proteins and COD and reduction of ionic concentration in dairy process waters (initial COD of 36,000 mg/L) using high shear rate dynamic filtration systems with NF (Birwal et al., 2017).

A two-stage ultrafiltration and nanofiltration (UF/NF) measure for the treatment of model dairy wastewater explored to reuse supplements and water from the wastewater. Ultracel PLGC and NF270 membranes were found to be the most suitable for this reason. In the primary stage, protein and lipid were concentrated by the Ultracel PLGC UF membrane. They could be used for algae cultivation to produce biodiesel and biofuel. The permeate from UF was concentrated by the NF270 membrane in the subsequent stage to obtain lactose in the retentate and reusable water in permeate.

In contrast, the NF retentate could be recycled for anaerobic digestion to produce biogas. With this methodology, most dairy wastewater could be recycled to produce reusable water and substrates for bioenergy production. Compared with the single NF process, this two-stage UF/NF process had a higher efficiency and less membrane fouling (Luo et al., 2011).

Treatment of Dairy Wastewater with a Membrane Bioreactor

The use of the aerobic membrane bioreactor (MBR) to treat dairy industry wastewater was shown to be feasible. Permeate of the MBR showed excellent quality with low concentrations of organic matter and nutrients. The efficiencies of the MBR working at hydraulic retention time (HRTs) of 6 and 8 hours were similar, so the HRT of 6 hours was picked as the ideal since its adoption would allow smaller reactors. It was possible to keep the system pressure stable even with permeate fluxes as high as 27.5 $L/h.m^2$, based on the successful air circulation applied among the strands. Nonetheless, there was an increment in fouling when the flux was raised to 37.5 $L/h.m^2$ (Andrade et al., 2013).

An elite bioreactor, an aerobic jet loop reactor, combined with a ceramic membrane filtration unit, was utilized to explore its suitability for dairy processing wastewater treatment. The bioreactor's oxygen transfer rates were found to be very high (100–285 h−1) on the operating conditions. A loading rate of 53 kg COD m−3 d−1 resulted in 97–98% COD removal efficiencies under 3 h hydraulic retention time. The high mixed liquor-suspended solid (MLSS) concentrations could be retained in the system (up to 38,000 mg/L) with the contribution of the UF (ultrafiltration) unit. During the filtration of activated sludge, the fluxes decreased with increasing MLSS. Cake formation fouling was determined as the dominant fouling mechanism. The results demonstrate that the jet loop membrane bioreactor system was a suitable and effective treatment choice for treating dairy industry wastewater (Farizoglu, & Uzuner, 2011).

Biological Treatment

In general, dairy wastewater is treated by biological methods such as activated sludge process, aerated lagoons, trickling filters, sequencing batch reactor, up-flow anaerobic sludge blanket reactor, anaerobic filters, etc. (Lateef et al., 2013). Biological treatment has the advantages of microbial changes of complex organic and possible adsorption of heavy metals by suitable microbes. These strategies have extraordinary potential for combining various types of biological schemes for particular constituent's removal. The presence or absence of oxygen may be further classified into two groups- aerobic and anaerobic.

Aerobic Process

Sequencing Batch Reactor (SBR)

Sequencing batch reactor (SBR) seems to be the most promising innovative technology in which dairy wastewater can be treated in a set of tanks that work based on a fill-and draw-activated sludge for eliminating undesirable components. These tanks might be an earthen or oxidation ditch, a rectangular basin, or any other concrete/ metal type structure. Each tank in the SBR system is filled during a discrete period and then worked as a batch reactor. After desired treatment, the blended liquor is allowed to settle, and the clarified supernatant is drawn from the tank. The fundamental contrast between the SBR and the conventional continuous-flow activated sludge system is that SBR carries out functions such as equalization, aeration, and sedimentation in a time rather than in a space sequence. The investment cost of SBR treatment is low because it is often operated with higher TS. The working parameters such as hydraulic retention time (HRT), organic loading, pH, dissolved oxygen concentration, temperature, mixed liquor-suspended solid (MLSS), and the strength of the wastewater affect the working treatment efficiency of SBR (Ahmad et al., 2019).

Anaerobic Process

The anaerobic process is a biological process performed by dynamic microorganisms without exogenous electron acceptors. It was accounted for that up to 95% of the natural burden in the wastewater stream can be changed to biogas (methane and carbon dioxide). Anaerobic process of dairy waste likewise yields methane, which can be utilized as a heat or power source. Another advantage of this process is that less sludge is created, thereby eliminating problems associated with sludge disposal, low area demand, no energy requirement, and absent bad scents. Anaerobic systems are generally seen as more prudent for the biological stabilization of dairy wastes, as they do not have the high-energy requirements related to aeration in aerobic systems.

Up Flow Anaerobic Sludge Blanket (UASB)

UASB reactors have been utilized in modern dairy wastewater treatment for over 20 years. They are appropriate for the treatment of overloaded effluents with COD higher than 42 g/L. Research center scale UASB reactors use whey permeates in a continuous regime have been planned. Kinetic coefficients using the Monod equations are resolved per HRT of 0.4–5 days and an underlying wastewater COD of 10.4–0.2 g/L. It was explained by a comparative study of the chance of utilizing flocculent sludge and the impact of different HRTs (6–16 h) on the anaerobic UASB reactor behavior applied to dairy wastewater treatment that nearly 80% of protein mineralization, soluble COD, and volatile fatty acid degradation as well as over 60% fat removal can be reached at an HRT of at least 12 h. During the utilization of industrial effluent from Edam cheese, butter, and milk production, a full-scale UASB reactor can be applied, the COD being diminished by 70% (Kolev, 2017).

Hybrid Anaerobic Digesters

A lab-scale on mesophilic hybrid anaerobic digester, comprised of UASB and an anaerobic filter design, was utilized for dairy effluent treatment. The anaerobic digester gave a methane yield of 0.354 m3 CH4/ kg COD removed at an HRT of 1.7 days. The treatment of acidic cheddar whey by a lab-scale up-flow hybrid reactor resulted in removal efficiencies of more than 95 percent, at two days of HRT and up to an OLR of about 11 kg COD/m3day.

Anaerobic Sequencing Batch Reactors (ASBR)

ASBRs are additionally utilized for getting high-quality dairy effluents. The lab-scale study on the ASBR system was done to remove COD and BOD5 rates of 62 and 75 percent, respectively, at a hydraulic retention time (HRT) of 6 h, at 58 °C, for a substrate containing non-fat dry milk. While in the temperature range between 5 and 20 °C, and at an HRT range between 24 and 6 h, soluble organic removal rates ranged between 75 and 90% for BOD5, and 62 and 90 percent for COD. In another lab-scale study, thermophilic ASBR and mesophilic ASBR systems gave volatile organic solids removal between 26-44 percent and 26-50 percent for dairy wastewater, respectively.

Nutritional Value of Dairy Effluents

Whey comprises around 45-50 percent of total milk solids, 70 percent of milk sugar (lactose), 70-90 percent minerals, 20 percent milk proteins, and almost all water-soluble vitamins present in milk. In the cheese-making process, water-soluble vitamins are passed into whey to a differing extent, ranging from 55-75 percent vitamin B6 and pantothenic acid; 40-70 percent vitamin B12; 80-90 percent of thiamine, nicotinic acid, folic acid, and ascorbic acid; and 70-80 percent of riboflavin and biotin present in milk. The highest amount of vitamin B12 is moved in whey in rennet coagulation compared to acid coagulation (Panghal et al., 2018).

Whey proteins are further divided into two subclasses: major proteins and minor proteins. Major whey proteins are composed of beta-lactoglobulin (b -Lg) 65 percent, alpha-lactalbumin (a -La) 25 percent, immunoglobulins (Ig), bovine serum albumin (BSA) and protease peptones and Minor whey proteins are composed of lactoferrin, lysosome, Glyco-macropeptide (GMP), Phospho-lipoproteins and lactoperoxidase. Whey proteins possess a high protein efficiency ratio (3.6), net protein use (95), and biological value (110), making them superior to soy protein, casein, wheat gluten, or beef and just next to egg protein. Whey protein constitutes all of the 20 amino acids, including all non-essential and essential amino acids such as Iso-leucine, Lysine, Phenylamine, Methionine, Threonine, Leucine, Histidine, Tryptophan, and Valine. Sulfur-containing amino acids, i.e., methionine and cysteine, are present in significant amounts (Macwan et al., 2016).

Whey is a by-product of the dairy industry, which for quite a long time was believed to be inconsequential and was either utilized as an animal feed or discarded as waste. In the latter case, disposal of whey is problematic due to several reasons. First, it's high BOD (the amount of O_2 in mg, required for the biological oxidization of the organic load per liter of whey, in five days), which is about 35.000 – 55.000mg O_2/liter. Considering that over 145.000.000 tons of whey are produced worldwide annually, the longing for new techniques to use whey can be valued (Tsakali et al., 2010).

During the manufacturing process of cheese, milk whey is produced as a liquid by-product. It addresses 85% - 90% of the original milk volume and holds 55% of the nutrients. Milk whey has proteins, lipids, soluble vitamins, minerals, and carbohydrates. 75% of the total whey solids is lactose. Lactose is one of the significant contaminating by-products because of its high biochemical oxygen demand (BOD) < 35,000 ppm and its chemical oxygen demand (COD) < 60,000 ppm. Suppose this industrial effluent is disposed of in the soil without earlier treatment. In that case, it can alter the soil's physical and chemical composition, decreasing crop yields and the accessibility of oxygen in the water. It has been assessed that 40,000 liters of untreated milk whey produce contamination comparable to that created day by day by a population of 250,000 people. Another by-product from the dairy industry is clarified butter sediment waste produced from the manufacturing of clarified or ghee butter. This butter waste is mostly made out of fats in fatty acids and a small number of proteins. It has been found that discarding milk whey in the water of rivers and lakes causes huge contamination issues because of the large amounts of nitrogen and phosphorus and high BOD (40 - 48,000 mg/L) and COD (89 - 95,000 mg/L) contained in this by-product. The conditions referenced above promote the process of eutrophication, causing excessive growth of microorganisms and aquatic plants (De Jesus et al., 2015).

Segmentation of Global Functional Dairy Market

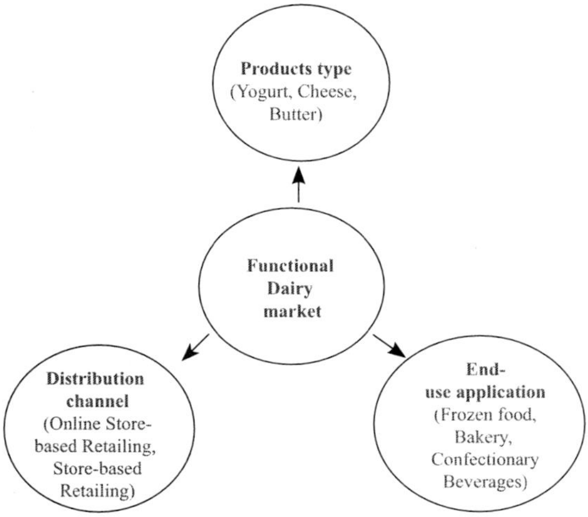

Fig. 4: Segmentation of functional dairy market at global perspective

Table 1: List of functional dairy products and their manufacturers (Source- Özer, & Kirmaci, 2010)

Functional dairy products	Bioactive compound	Manufacturer	Brand
Hyper-immune milk powder	Casein-phospho-peptide (CPP) & IgG	Stolle Milk Biologics	Stolle Milk Alpha
Fermented dairy – fruit beverage	Peptides obtained by milk fermentation	Valio, Finland	Evolus
Milk fruit juice drink	Conjugated linoleic acid (Tonalin brand from Cognis)	Corporation Alimentaria Penananta S.A. Spain	Natura Linea
Fermented milk drink	Magnesium	Danone Belgium	Zen
Caseino-phosphopeptide (CPP) and IgG rich milk	Enriched with CPP and IgG	Stolle Milk Japan	Alpha
Probiotic fermented milk	*Lb. casei*, *lactoferrin*	Kyodo Milk Japan	*Lb. casei* + lactoferrin
Cultured Milk	Bioactive peptides	Calpis Japan	AmealS

Utilization of Dairy Effluents in Development of Different Immune Boosting Product to Combat Covid 19 Pandemics

Preparation of whey-based papaya RTS - Whey can be utilized as a suitable ingredient to produce many healthy beverages and products like whey-based papaya RTS. It was found that the addition of 25% whey is most acceptable in terms of nutritional and consumer acceptability. This RTS contains 15.05±0.18

TSS, 0.30±0.11 % acidity, 5.37±0.01 g/100g reducing sugar, 14.06±0.46 g/100g total sugar, and 5.60±0.02 mg/100g ascorbic acid, thus depicting improved nutritional value and this creates a scope for better health, enhanced postharvest value, and whey utilization (Panghal et al., 2017).

Preparation of B12 enriched fermented whey - Whey samples had lactose (2·80-6·11%), protein (0·325-0·485%), fat (0·10-0·25%), and lactic acid (0·14-0·40%) contents resulting in a high Biochemical Oxygen Demand (BOD5 24·8–59·1 × 103 mg litre−1). It was analyzed that.*Propionibacterium shermanii*.566 synthesized 5·76 mg vitamin B12 per liter of whey containing 4% lactose supplemented with 0·5% (NH4)2HPO4 when fermentation was carried out at 30°C under anaerobiosis for the first half (84 h) followed by aerobiosis for the second half of the fermentation (84 h). It was observed that metabolite started accumulating at the end of the maximum growth phase (4th day) and lasted until the curve reached a plateau (7th day). Growth of.*P. shermanii*.566 also helps in the reduction of BOD of the whey by >90%, thereby decreasing the BOD of the milk plant effluents, and vitamin B12 enriched fermented whey was found to replace the animal protein factor (APF) as a vitamin B12 source in poultry rations. Therefore, the technology developed is a worthwhile attempt to utilize whey for vitamin B12 production and, in turn, reduce the water pollution problem (Marwaha and Shetti 1984).

Whey protein is a significant source of amino acids and is available in various forms like whey protein concentrate, Whey protein isolate, whey protein hydrolysate, and native whey (Sparsh and Parag 2017).

Utilization of whey in the production of extracellular protease- Dairy industry waste like fresh cheese whey has been utilized as a sole substrate for the production of extracellular protease by a locally isolated strain of.*Rhizopus oryzae*. In pure whey medium only mineral salts were added as (g/L): MnSO4. H2O (0.01%), FeSO4.H2O (0.01%), CaC12 (0.05%) at pH 6. In the submerged fermentation process, the effects of shaking speed, pH, temperature, and incubation time for protease production were studied. It was found that maximum production of extracellular alkaline protease was achieved after 168 h at 35°C. ANOVA explained that the mutant strain enzyme was novel, thermo-tolerant, and active at pH 10, resulting in 3.85 fold elevated enzyme activity (2538.9 PU/ml) under optimum fermentation conditions (Gul et al., 2012).

Role of immune milk in immunity-boosting. Notably, bovine colostrum and cow's milk are profoundly wealthy in IgG that ties to numerous gastrointestinal and respiratory microbes that infect humans, supporting cross-species activity between bovine and human IgG. Cow's milk is available to the whole population, and drinking micro-filtered raw immune milk from cows vaccinated against SARS-CoV-2 (example, SARS-CoV-2 S-glycoproteins) could give individuals with short-term protection against the SARS-CoV-2 infection until vaccines become commercially accessible. Immunized cows can produce specific IgG antibodies against SARS-CoV-2 glycoproteins in milk or colostrum. The microfiltration process eliminates all pathogens from the milk without denaturation of proteins, or nutritional changes brought about by the ultra-high temperature pasteurization of milk. IgG from non-immunized cows can interact with various kinds of pathogens, including viruses.

Interestingly, bovine IgG can interact with human Fcγ receptors (FcgR), supporting the idea that this interaction can enhance the presentation of antigens to T-cells and phagocytosis by leukocytes. A trial study explained that bovine colostrum increased the proportion of CD8+ T cells post-viral challenge in mice. IgG also has different functions, like the agglutination of pathogens, complement fixation for pathogen lysis, inhibition of pathogen metabolism by blocking enzymes, and neutralizing viruses. Bovine-colostrum-derived IgG can hinder the NF-κB signaling pathway and the production of pro-inflammatory cytokines in intestinal cells (Jawahara, 2020).

Utilization of dairy waste by different biotechnological methods- Biotechnological methods act as an elective way to deal with these dairy wastes in the production of biomass, bioplastics, bio-fertilizers, bio-surfactant, biofuels, single-cell protein formation, organic acid, and exopolysaccharides production (De Jesus et al., 2015). However, dairy wastage can be utilized to deliver the immune-boosting product to combat Coronavirus disease. Many products can also be reformulated using these dairy wastes with turmeric, tulsi, mulethi, ashwagandha and some other herbs to boost immunity.

Preparation of hydrolyzed collagen added whey-based beverages- Milk whey-based beverage with various concentrations of hydrolyzed collagen (0.3%, 0.5%, 0.75%, and 1%) was prepared. It was compared with the control sample. Their physicochemical properties, viscosity, antioxidant activity, and microbiological parameters were assessed. It was found that 1% collagen treatment indicated the highest protein content (9.75 ± 0.20 g/L), as well as radical inhibition for ATBS (48.30%) and DPPH (30.06%). There were no significant differences ($p \geq 0.05$) in the fat and lactose parameters. However, the pH in the control treatment was lower as compared to beverages treated with hydrolyzed collagen. The viscosity increased significantly as the concentration of hydrolyzed collagen increased. This experiment showed that hydrolyzed collagen›s addition increased the bioavailability, nutritional value, and antioxidant activity of the beverage. Hydrolyzed collagen acted as an antimicrobial agent, as there was no presence of microorganism pathogens observed in the treated beverages (León-López et al., 2020).

Formulation of Fermented drink- This work centers around the preparation of liquid whey protein concentrate by ultrafiltration followed by thermal denaturation and homogenization of the ultrafiltrated concentrate, and the production of ultrafiltrated permeates concentrated by reverse osmosis. Kefir grains (fresh and thawed) and/or commercial probiotic bacteria were inoculated in both liquid whey protein concentrates and concentrated ultrafiltrated permeates and grown at 25 °C for 24 h for the manufacture of fermented drinks. The physicochemical characterization (pH, titratable acidity, viscosity, and content of total solids, ash, fat, and proteins) of the obtained drinks was analyzed and compared. Enumeration of viable microorganisms was carried out immediately after inoculation (at 0 h), during the fermentation period (at 12 and 24 h), and during refrigerated storage (at 48, 168, and 336 h). It was found that the fermented drinks possess acceptable physicochemical and sensorial properties and contained above 7 log CFU/mL of lactococci and lactobacilli and 6 log CFU/mL of yeasts after 14 days of refrigerated storage, which is in concurrence with the norms needed by international organizations like European Food Safety Authority (EFSA) and Food and Drug Administration (FDA) for products containing probiotics (Pereira et al., 2015).

Table 2: List of dairy products, their by-products and encapsulation technique

Dairy Product	Dairy Byproduct	Encapsulation technique	Reference
Fermented milk	Whey Protein	-	Zhang et al., 2015
Yogurt	Whey Protein Concentrate	-	Sandoval-Castilla et al., 2004
Cheese powder	Fluid Whey	-	Erbay et al., 2015
Peptide-enriched powder	Buttermilk	Spray drying	Burns et al., 2010
Galacto-oligosaccharides.	Cheese Whey Permeate	Freeze drying	Golowczyc et al., 2013
Kefir	Cheese Whey	-	Londero et al., 2011

Production of Whey proteins hydrolysates by using trypsin enzyme- Whey is a protein complex derived from milk, exhibit most protein quality rating among other proteins, being promoted as a functional food with several medical advantages. In this experiment, whey protein, protein hydrolysates are produced using trypsin enzymes to augment antioxidant activity and assess angiotensin-converting enzyme (ACE) inhibition activity. Hydrolysis parameters were standardized applying response surface methodology. The assay method's radical scavenging assay method determined the response antioxidant activity in Trolox equivalent antioxidant capacity (TEAC) values.

Optimum conditions for maximum antioxidant activity were standardized at 88 °C of preheating, 7.3 pH, 0.05 enzymes to substrate ratio, and hydrolysis were carried up to 8 h at 36.5 C. Resulting peptide fractions obtained at 11.8% of the degree of hydrolysis displayed antioxidant capacity with TEAC values of 1.37 ± 0.12. The designed model obtained to be significant with an R2 value of 0.9972 for antioxidant activity and lack of fit test-as non-significant, indicating that the optimized conditions were best suited. The hydrolysate was further investigated for antihypertensive activity. The results indicate that to affect the decrease in ACE inhibition activity, 4,166.72 lg of native whey protein is required compared to 229.96 lg of hydrolysates. These results indicate hydrolysate produced under these conditions could be an effective nutraceutical (Naik et al., 2013). Thus, various value-added products could be formulated by using dairy industry waste.

Conclusion

The outbreak of coronavirus has had a significant impact on public health worldwide and various industries' supply chain and value chain stages. This disease's typical symptoms include fever, cough, myalgia, or fatigue, and severe complications include acute respiratory distress syndrome (ARDS), RNAemia, acute cardiac injury, and multiple organ failure. However, exaggerated immune responses may lead to Cytokine Storm. The global impact of this new epidemic is yet uncertain. So, it is advised to boost your immunity, which can be possible by adopting changes in diet like supplementation of immune-boosting products. At the same time, dairy waste management is a significant concern to reduce the pollutant loads and its utilization in the formulation of immune booster product development because dairy waste has potent ingredients to utilize in various forms like product formulation. However, these wastes could efficiently develop diversified value-added functional dairy foods such as health drinks and beverages. These may also serve as a potential source of vitamin B12, protease production, hydrolyzed collagen-based beverages, fermented drinks, and many more. There is an urgent need to conduct extensive research to utilize such dairy waste to develop diversified functional food with immense potential to combat chronic diseases like cancer, neurological disorders, diabetes, hypertension, autoimmune disorders, and deficiency syndromes. In the era covid-19 pandemic, this may serve as an efficient tool to boost the immune system.

References

Ahmad, T., Aadil, R. M., Ahmed, H., ur Rahman, U., Soares, B. C., Souza, S. L., and Freitas, M. Q. (2019). Treatment and utilization of dairy industrial waste: A review.Trends in Food Science & Technology,.88, 361-372.

Andrade, L. H., Motta, G. E., & Amaral, M. C. S. (2013). Treatment of dairy wastewater with a membrane bioreactor.Brazilian Journal of Chemical Engineering,.30(4), 759-770.

Arshad, M. S., Khan, U., Sadiq, A., Khalid, W., Hussain, M., Yasmeen, A. and Rehana, H. (2020). Coronavirus disease (COVID-19) and immunity booster green foods: A mini review.Food Science & Nutrition, 8(8): 3971-3976.

Birwal, P., Deshmukh, G., Priyanka, S. S., and Saurabh, S. P. (2017). Advanced technologies for dairy effluent treatment.J Food Nutr Popul Health,.1(1), 7.

Burns, P., Molinari, F., Beccaria, A., Paez, R., Meinardi, C., Reinheimer, J., and Vinderola, G. (2010). Suitability of buttermilk for fermentation with Lactobacillus helveticus and production of a functional peptide-enriched powder by spray-drying.Journal of applied microbiology,.109(4), 1370-1378.

Chezeau, B., Boudriche, L., Vial, C. and Boudjemaa, A. (2020). Treatment of dairy wastewater by electrocoagulation process: Advantages of combined iron/aluminum electrodes. Separation Science and Technology,.55(14), 2510-2527.

Das, B., Sarkar, S., Sarkar, A., Bhattacharjee, S. and Bhattacharjee, C. (2016). Recovery of whey proteins and lactose from dairy waste: A step towards green waste management. Process Safety and Environmental Protection,.101, 27-33.

Das, M., Raychaudhuri, A. and Ghosh, S. K. (2016). Supply chain of bioethanol production from whey: A review.Procedia Environmental Sciences,.35, 833-846.

de Faria Coelho-Ravagnani, C., Corgosinho, F. C., Sanches, F. L. F. Z., Prado, C. M. M., Laviano, A. and Mota, J. F. (2020). Dietary recommendations during the COVID-19 pandemic.Nutrition Reviews.

De Jesus, C. S. A., Ruth, V. G. E., Daniel, S. F. R., and Sharma, A. (2015). Biotechnological alternatives for the utilization of dairy industry waste products.Advances in Bioscience and Biotechnology,.6(03), 223.

Devi, M. G., Dumaran, J. J., and Feroz, S. (2012). Dairy wastewater treatment using low molecular weight crab shell chitosan.Journal of The Institution of Engineers (India): Series E,.93(1), 9-14.

Erbay, Z.; Koca, N. (2015). Effects of whey or maltodextrin addition during production on physical quality of white cheese powder during storage. J. Dairy Sci., 98, 8391–8404.

Farizoglu, B. and Uzuner, S. (2011). The investigation of dairy industry wastewater treatment in a biological high performance membrane system.Biochemical Engineering Journal,.57, 46-54.

Golowczyc, M., Vera, C., Santos, M., Guerrero, C., Carasi, P., Illanes, A. and Tymczyszyn, E. (2013). Use of whey permeate containing in situ synthesised galacto-oligosaccharides for the growth and preservation of Lactobacillus plantarum. The Journal of dairy research,.80(3), 374.

Gopinathan, M., and Thirumurthy, M. (2012). Feasibility Studies on Static Pile Co Composting of Organic Fraction of Municipal Solid Waste With Dairy Waste Water.Environmental Research, Engineering and Management,.60(2), 34-39.

Gordon, C. J., Tchesnokov, E. P., Feng, J. Y., Porter, D. P., and Götte, M. (2020). The antiviral compound remdesivir potently inhibits RNA-dependent RNA polymerase from Middle East respiratory syndrome coronavirus.Journal of Biological Chemistry, 295 (15), 4773-4779.

http://www.fao.org/dairy-production products/production/en/#:~:text=India%20is%20the%20 world's%20largest,%2C%20China%2C%20Pakistan%20and%20Brazil.).

Jawhara, S. (2020). Can drinking microfiltered raw immune milk from cows immunized against SARS-CoV-2 provide short-term protection against COVID-19?.Frontiers in immunology, 11, 1888.

Kolev Slavov, A. (2017). General characteristics and treatment possibilities of dairy wastewater–a review.Food technology and biotechnology, 55(1), 14-28.

Kothari, R., Pathak, V. V., Kumar, V. and Singh, D. P. (2012). Experimental study for growth potential of unicellular alga Chlorella pyrenoidosa on dairy waste water: an integrated approach for treatment and biofuel production.Bioresource technology,.116, 466-470.

Kurup, G. G., Adhikari, B., and Zisu, B. (2019). Recovery of proteins and lipids from dairy wastewater using food grade sodium lignosulphonate.Water Resources and Industry,.22, 100114.

Lagrange, V., Whitsett, D. and Burris, C. (2015). Global market for dairy proteins.Journal of food science, 80(S1), A16-A22.

Lateef, A., Chaudhry, M. N., and Ilyas, S. (2013). Biological treatment of dairy wastewater using activated sludge.Science Asia,.39(2), 179-185.

León-López, A., Pérez-Marroquín, X. A., Campos-Lozada, G., Campos-Montiel, R. G., and Aguirre-Álvarez, G. (2020). Characterization of Whey-Based Fermented Beverages Supplemented with Hydrolyzed Collagen: Antioxidant Activity and Bioavailability.Foods, 9(8), 1106.

Londero, A., Quinta, R., Abraham, A. G., Sereno, R., De Antoni, G., & Garrote, G. L. (2011). Inhibitory activity of cheese whey fermented with kefir grains.Journal of Food Protection, 74 (1), 94-100.

Luo, J., Ding, L., Qi, B., Jaffrin, M. Y., & Wan, Y. (2011). A two-stage ultrafiltration and nanofiltration process for recycling dairy wastewater.Bioresource Technology,102(16), 7437-7442.

Marwaha, S. S. and Sethi, R. P. (1984). Utilization of dairy waste for vitamin B12 fermentation. Agricultural wastes,.9(2), 111-130. P., Srivastava, V. C., & Mall, I. D. (2010). Organics removal from dairy wastewater by electrochemical treatment and residue disposal. Separation and Purification Technology,.76(2), 198-205.

Mohammed, A. S. (2017). Electrochemical and Electroflotation Processes for Milk Waste Water Treatment.(Doctoral dissertation, Cleveland State University).

Moradi, O. and Maleki, M.S. (2013). Removal of COD from dairy wastewater by MWCNTs: adsorption isotherm modeling.Fullerenes, Nanotubes and Carbon Nanostructures,.21(10), 836-848.

Naik, L., Mann, B., Bajaj, R., Sangwan, R. B. and Sharma, R. (2013). Process optimization for the production of bio-functional whey protein hydrolysates: adopting response surface methodology.International Journal of Peptide Research and Therapeutics, 19(3), 231-237.

Özer, B. H. and Kirmaci, H.A. (2010). Functional milks and dairy beverages. International Journal of Dairy Technology,.63(1), 1-15.

Panghal, A., Kumar, V., Dhull, S. B., Gat, Y. and Chhikara, N. (2017). Utilization of dairy industry waste-whey in formulation of papaya RTS beverage.Current Research in Nutrition and Food Science Journal,.5(2), 168-174.

Panghal, A., Patidar, R., Jaglan, S., Chhikara, N., Khatkar, S. K., Gat, Y. and Sindhu, N. (2018). Whey valorization: current options and future scenario–a critical review.Nutrition & Food Science.

Pereira, C., Henriques, M., Gomes, D., Gomez-Zavaglia, A. and de Antoni, G. (2015). Novel functional whey-based drinks with great potential in the dairy industry. Food technology and biotechnology,.53(3), 307-314.

Rad, S. J., and Lewis, M. J. (2014). Water utilisation, energy utilisation and waste water management in the dairy industry: A review.International Journal of Dairy Technology, 67(1), 1-20.

Rani, R. U., Kumar, S. A., Kaliappan, S., Yeom, I., & Banu, J. R. (2013). Impacts of microwave pretreatments on the semi-continuous anaerobic digestion of dairy waste activated sludge. Waste Management,.33(5), 1119-1127.

Rao, M., and Bhole, A. G. (2002). Removal of organic matter from dairy industry wastewater using low-cost adsorbents. Indian Chemical Engineer,.44(1), 25-28.

Sandoval-Castilla, O., Lobato-Calleros, C., Aguirre-Mandujano, E., Vernon-Carter, E.J. Microstructure and texture of yogurt as influenced by fat replacers. Int. Dairy J. 2004, 14, 151–159.

Singhal, T. (2020). A review of coronavirus disease-2019 (COVID-19).The Indian Journal of Pediatrics, 1-6.

Suman, A., Ahmad, T., and Ahmad, K. (2018). Dairy wastewater treatment using water treatment sludge as coagulant: a novel treatment approach.Environment, Development and Sustainability, 20(4), 1615-1625.

Tchamango, S., Nanseu-Njiki, C.P., Ngameni, E., Hadjiev, D., and Darchen, A. (2010). Treatment of dairy effluents by electrocoagulation using aluminium electrodes.Science of the total environment, 408(4), 947-952.

Tsakali, E., Petrotos, K., D'Allessandro, A., and Goulas, P. (2010, June). A review on whey composition and the methods used for its utilization for food and pharmaceutical products. In.Proc. 6th Int. Conf. Simul. Modelling Food Bioind.(pp.195-201).

Vibhhute, H. V., and Ingavale, B. (2020). Effect of F/M Ratio on the Performance of Sequencing Batch Reactor for Dairy Wastewater Treatment.Aquademia,.4(2), ep20021.

Zhang, T., Mccarthy, J., Wang, G., Liu, Y., Guo, M. (2015) Physiochemical properties, microstructure, and probiotic survivability of nonfat goats' milk yogurt using heat-treated whey protein concentrate as fat replacer. J. Food Sci. 80, M788–M794.

Zheng, S., Baak, J.P., Li,S., Xiao, W., Ren, H., Yang, H. and Wen, C. (2020). Network pharmacology analysis of the therapeutic mechanisms of the traditional Chinese herbal formula Lian Hua Qing Wen in Corona virus disease 2019 (COVID-19), gives fundamental support to the clinical use of LHQW. Phytomedicine, 79: 153336.

17

Packaging of Functional Foods and Nutraceuticals

Priyanka Sengupta **and** ***Ayan Dey***

Indian Institute of Packaging, E2- Andheri East, Mumbai-400093, Maharashtra

Introduction

Food and nutraceuticals are obtained from various agricultural resources available in this planet. Many types of bio-active compounds are available in small quantities in the plants and animals which are evidenced to improve performance of various organs, increase in immunity and cure for some diseases. Last decade evidences several advancements in developing phyto-chemical formulations which can be used as a cure for many diseases as well as helps in boosting immunity in human being. Such materials exist in different form, structure and chemical compositions. Examples of such chemicals are phenolic and organosulphur compound (Pascall 2011). The vegetables, fruits and seeds are used without further processing. Some plants are subjected to cooking before its consumption. Many industries are carried out extraction of bioactive compounds and developed products in the form of tablets, powders and liquids which are available in the market. In order to make commercial process more economic and feasible, long term storage of plant residues (Raw materials) as well as the prevention of the extracts emerges as essential requirement for the industries. Moreover, the products requires suitable packaging solution to increase its self-life, preventing it from the influences of environment, and to make viable solution for easy access which makes it more user-friendly. Suitable packaging solution is required for mainly the three reasons: (i) prevention. (ii) protection and (iii) presentation. Prevention is required in-order to avoid deterioration of the packaged items due to microbial growth, loss in moisture and enzymatic action. Protection is required for the purpose of protecting the packaging materials from the contamination as well as from the environment impact and rough handling. Presentation is equally important to provide it an aesthetic look to make it more approachable to the consumers. Moreover, packaging also offers "labelling for easy identification, warnings, instructions, and nutritional information, ease of transportation" (Pascall 2013). It is also noteworthy to mention here that Anne-Sophie Binninger (2015) divided the purpose of the packaging into three groups which are: (i) commercial functions, (ii) physical functions and (iii) social functions. The commercial functions can be achieved through easy "identification, communication, positioning, and distinction" where as the physical functions are storing and providing protection as well as the conservation of the contents. The social functions of packaging are to reduce pollution and to encourage recycling of the packaging materials (Binninger 2017).

In the subsequent sections, packaging solutions for various foods, nutraceutical and bio-active compounds and the material types used in developing packages are discussed briefly. Pre- and probiotics, synbiotics, dried food, fermented products, fresh fruits and vegetables are the examples of functional food items. The bioactive components present in it are "minerals, vitamins, fibers, peptides, proteins, n-3 polyunsaturated fats, antioxidants, enzymes, symbiotics and other micronutrients" (Deschênes 2015).

Factors causing spoilage of functional food items

The most common factors which are the major reason for the spoilage of food items are (i) Microorganism, (ii) Enzymes, (iii) Air, (iv) Light, (v) Insects, Rodents, Parasites etc., (vi) Physical Damage and (vii) Temperature. The microorganisms found to be present in food items are there types, which are Bacteria, Protozoa and fungi. The growth of the most of the microorganisms is facilitated at a temperature of 60-90 °F. However, some bacteria, yeast, molds can grow at a low temperature (40°F). Enzymes, on the other hand present in the food and is the reason of ripening process. Such phenomenon causes the change in taste, colour, flavour and texture of the food items. This is the reason of changing colour of banana from green to yellow and finally to brown. Food contamination results various food-borne diseases which are affecting human health every year (Kamala and Kumar 2018). Functional ingredients are evidenced to be susceptible to oxygen and moisture (Bell 2006). Hence, the packaging material that can able to act moisture and oxygen barrier, is able to greatly improve the shelf-life of the packaged functional ingredients. The shelf-life can be further enhanced by modified atmosphere packaging and antimicrobial packaging technologies which are discussed in the subsequent section.

Factors that affect the stability of the nutraceuticals are described by Gupta et al. (2019). These factors are (i) storage conditions, (ii) drug interactions, (iii) content variations, (iv) Packaging interactions, (v) container characteristics,

(vi) moisture content, (viii) mold growth, (viii) microorganisms, (ix) insect attack and (x) various environmental factors (Gupta et al. 2019). According to Gupta et al. (2019), "A nutraceutical product may undergo change in appearance, consistency, content uniformity, clarity (solution), moisture content, particle size and shape, pH, and package integrity, thereby affecting its stability" (Gupta et al. 2019).

Table 1: The phenomena and its effect on nutraceutical and phyto-formulations (Gupta et al. 2019)

Phenomenon	Causes
Chemical reactions like solvolysis, oxidation, reduction, racemization, etc.	Degradation products, loss of excipient activity, potency loss
Enzymatic deterioration	Degradation
Growth of microorganisms	change in appearance, consistency, clarity, pH, favour and taste, preservative efficacy
Light	Affects phyto-formulations by generating radical
Moisture	Increases the rate of degradation
environmental factors of oxygen and carbon dioxide	Affects stability of phyto-formulations
Compositions like water, other solvents, emulsifier etc.	Affects stability of phyto-formulations
Contaminations from containers	Affects stability of phyto-formulations and nutraceauticals

Oyetade et al. (2012) reported a decrease in Ascorbic acid content significantly even at a temperature of 4 °C (The Ascorbic acid content in grape juice decreases from 58.2 to 45.8 in 4 h) (Oyetade e al., 2012).

Various functional foods and nutraceuticals and its packaging practices

Fruits and vegetables

Fruit and vegetables are rich source of phenolic compounds which gains interest due to its antioxidant characteristics (Shahidi and Naczk 2004). However, such molecules are very susceptible towards oxygen and readily oxidized resulting loss in its activity. Phenolic compounds are also found in several plant products like "cereals, legumes, oilseeds, fruits, and plant origin beverages" (Deschênes 2015). Moreover, plant tissues are made of watery solutions or dispersions which are readily influenced by any kind of physical, chemical or enzymatic reactions. According to Venir and Maltini (2013), the physical properties of the food items are affected by water however such effect depends on the interaction between water and the chemical constituents of the food (Venir and Maltini 2013). Processing and storage conditions greatly affect the stability of the antioxidants and hence it is essential to control the oxygen level in order to enhance the shelf life of the food product and to maintain the "Naturalness" of the packaged food items. High temperature often can be used but it can be detrimental to some food items. The high temperature processing consists of several steps which are blanching, pasteurization and sterilization and thermal drying. Blanching is the process of inactivation of the enzymes maintaining color and nutrition of the food items. Several mediums are proposed for the use in blanching which are water, steam, vacuum steam, hot air etc. Among them water is more popular as it makes the process economic. Water blanching is carried out at a temperature of 75-95 °C for 1-10 min. However, Pacheco-Palencia et al. (2009) pointed out a drawback of this process. Being a thermal process it results in loss of bio-active compounds like anthocyanins which is found in fruit puree of acai fruit species, pineapple, papaya cubes etc. It is a heat sensitive element and so degrades fast when exposed to a temperature of 80 °C for 1 to 60 min. (Pacheco-Palencia et al. 2009. Moreover, it is also established that ascorbic acid is thermally labile element which can easily degrades at high temperature. (Beyers and Thomas 1979. Blanching is considered as a pretreatment stage. The next step is high temperature pasteurization and sterilization. The process results in destruction of microorganisms and inactivation of the enzyme pectin methylesterase. However, this process was found to deactivate and degrade the bio-active components like vitamin A, ascorbic acid, carotenoid etc. (Rattanathanalerk et al., 2009; Zepka & Mercadante, 2009; Aramwit et al., 2010; Chin et al., 2010;). In the next subsequent step, the drying process is used for volume reduction and for the extension of shelf-life as (Prakash et al. 2004). According to Rawson et al. (2011), several dehydration processes are used for drying fruits and vegetables which are "sun drying, cross-flow, fluidized bed, osmotic air drying, oven drying, drum, spray, puff, freeze and microwave drying" (Rawson et al. 2011). Though drying increases the shelf life of the fruits still it affects the presence and stability of bioactive compounds such as ascorbic acid, polyphenols, and carotenoids due to their sensitivity towards heat. Rawson et al. (2011) reported several non-thermal processing techniques like Dense phase carbon dioxide (DPCD), Pulsed electric field (PEF), ozone processing, ultrasound processing and high hydrostatic pressure (HHP) processing techniques as a better option to preserve the quality of bio-active ingredients present in fruits and vegetables(Rawson et al., 2011). However in order to resist the enzymatic degradation of bioactive compounds present in fruit and vegetables, mild heat treatment is considered as a very popular strategy to destroy the enzyme which promotes such degradation e.g. polyphenol oxidase enzyme is deactivated by using a heat treatment at 50 °C which exhibits a positive influence on anthocyanin (Rawson et al., 2011).

Edible coating materials are used to coat the fruits and vegetables containing bio-active ingredients in order to prevent its nutrition loss as well as increasing its shelf-life. Dip coating and drying technique are used to form coating over the fruit items (Raybaudi et al. 2010).

Prebiotics, Probiotic and Synbiotics

Probiotics bacteria are evidenced to be beneficial to human health. The food ingredients, type of packaging material and storage environment have a combined effect on the survival of probiotics. It is a general practice to use growth promoters like glucose, vitamins, proteins, minerals, caseins, yeast extract, antioxidant etc. (Mizielińska et al. 2017). Oxygen is evidenced to affect the probiotic by (i) direct poison the cells, (ii) producing toxic peroxides and (iii) generating free radicals which affect the probiotic cell (Tripathi et al. 2014; Korbekandi et al. 2011). High barrier film and the glass are proved to be effective packaging materials than the using many additives which exhibits a positive influence on the survival of probiotic bacteria for longer period (Tripathi et. al. 2014).Prebiotics can be defined as "a substrate that is selectively utilized by host microorganisms conferring a health benefit" (González-Herrera et al. 2020).Benefits of such bio-active compound are pathogens inhibition, immuneresponse stimulation, lipid levels reduction in blood, improving the mental health, bones etc. Prebiotics are established as carbohydrate based mainly. However, there are other evidences for other substances like "polyphenols and polyunsaturated fatty acids converted to their respective conjugated fatty acids" which can function as host (González-Herrera et al. 2020). Synbiotic is actually a combination of probiotics and prebiotics. According to Bandyopadhyay and Mandal 2014, synbiorics are more beneficial for the health compared to the effect of probiotic or prebiotics when used as individuals (Bandyopadhyay and Mandal 2014). The level of oxygen in the package is an important factor which determines the shelf-life of the packaged items. The pH of the medium is an important factor which is evidenced to affect the viability of probiotics. In case of ice cream made from goat milk, pH is at higher side if the milk is subjected to short fermentation time (1h). It is recommended to maintain pH between 5 and 5.5 in order to inhibit viability loss of the probiotics present in ice cream (Senaka Ranadheera et al. 2013).According to Saint-Eve et al. (2008), "the fat content of yogurt can act as an aroma solvent and reduce absorption into packaging. High fat content (~10%) can eliminate the influence of packaging materials like glass, polypropylene, polystyrene on the aroma and taste of the probiotic based food items (Saint-Eve et al. 2008).Functional food and nutraceuticals are also processed to obtain various products in the form of granules, powders, cream and gels.

Granules and powders

Bioactive plants are generally harvested and dried. The convenient method of storing dried bioactive plants is in the form of "powders, pills, granules, or in a partially crushed state". This not only reduces volume but makes them more convenient for transportation and storage. The plants parts like leaves, stems, barks, roots, seeds, fruits, and flowers are sectioned and then crushed, chopped, or ground for size reduction. Drying (procedure to remove moisture progressively) is done before the size reduction process in some cases and in some cased drying is carried out after the size reduction procedure. On properly drying, enzymatic activities, bacterial and fungal growth are significantly reduced. The product being very moisture sensitive and highly hydrophilic (in some cases) requires high moisture barrier characteristic of the packaging materials. Metal and glass container can be the best fit as packaging materials but there are some serious drawbacks of such materials. For these types of materials, leaks can be developed "if the closure is defective or not properly applied, or is incompatible with the sealing surface of the main package" (Pascall 2011). Many times partial consumption of product is observed after opening the package. Hence, the requirement of resealable container which permits the container to be resealed multiple times, are endorsed for the packaging materials. A sachet filled with desiccant in the package is also used to reduce moisture content in the headspace of the package before or after opening by a consumer. Modified atmosphere packaging is also used as an additional remedy in order to maintain a desired atmosphere within the closed packages. According to Pascall (2011), "if too much moisture enters the package, it can cause caking of powders, swelling and cracking of pills, loss in texture to baked products, reoccurrence of enzymatic activities within the plant tissue, and growth of microorganisms" (Pascall 2011).Plant bioactive compound is marketed in the form of small pellets, tablets, or pills coated with an edible additive (Gil et al. 2008). When "two-piece capsule" is used then the outer part is made of edible plastic materials which are mainly composed of starch or gelatin.

Creams and gels

Creams containing plant bioactive compounds are composed of mainly vegetable oils and glycerin with beeswax. A tincture of benzoin is used as preservative (Pamplona-Roger 1998). In some cases, vitamin E is also used as an antioxidant. Lecithin is used as an emulsifying agent to make the product homogeneous in nature and prevent separation of the water from the oils. A blend of bioactive ingredient extracted from plant and the aforesaid mixture is used to formulate the creams. The main storage problems with creams are oxidation and resultant rancidity of the product and microbial contamination. Oxidation causes the development of off-odors and a lossin the consistency of the product. As creams are mainly for external use, flavor is not a considerable issue that requires much attention. An inert gas is injected into the headspace in order to minimize the extent of oxidation of the packaged product (Mahanjan et al. 2005). This inert gas may be nitrogen or a mixture of nitrogen and carbon dioxide. To prevent bacterial contamination, pinhole free bottles and jars are used to package these products and it should be equipped with resealable devices. Tamper evident devices are essential in order to prevent pilferage in stores. Shrink bands, pull-tabs are the examples of temper evident device which are used in packing nutraceuticals. Many bioactive compounds can be used in the form of gels. Various hair products, lip balms, and deodorants are the examples of this kind. Gels are semi-crosslinked viscoelastic materials that are marketed in the form of liquids, emulsions and suspensions. Gelling agents are used to thicken and stabilize the liquid phase and form a colloidal mixture. Collapsible tube and plastic bottles with various types of dispensers are used to package creams and gels. Oxidation, microbial contamination, and deactivation of the bioactive ingredients present in the gels are appeared as major problem faced by the manufacturers. Moreover, these products are found to lose consistency if stored in conditions of high temperature or if subjected to severe vibrations over time. Because gels are high moisture products, they can lose moisture with time and become hard. Thus, the packaging types to be used for this type of products require protection "against light, heat, moisture loss, oxygen ingression, and microbial contamination" (Pascall 2011). Gels are difficult to dispense and hence for the convenience and ease to use, gels are usually packaged in "jars, squeezable bottles, disposable pouches, or seamless collapsible tubes" (Pascall 2011). Pouches are made of plastic films, whereas both aluminum and plastic are used to make tubes.

Packaging materials and types

Different types of packaging technologies are available for packing functional food items in order to provide protection, extend shelf-life and for improving aesthetics of the materials. Flexible packaging materials and rigid packaging materials are the two types of packages are used for packing functional food items. Conventional thermoplastic polymers like polyethylene, polypropylene, polystyrene, poly (ethylene terephthalate), Nylon 6, polylactic acid etc. are used to make packages. In order to meet demand of higher shelf life and durability, barrier property, it is general practice to develop either laminates or composites or blends using the thermoplastic polymers as matrix. Last decade has evidenced many developments in packaging for such food items among them antimicrobial packaging, modified atmosphere packaging, bioactive packaging and smart packaging etc. In the present scenario, the packaging technology is more inclined towards more eco-friendly packaging solutions which are either completely recyclable or biodegradable and biocompatible in nature. Many packaging materials are developed which are commercially used and 100% biodegradable or bio-compostable in certain condition. Among them, polylactic acid gains popularity due to its easy processing, durability and good barrier characteristics. Research in edible packaging materials is also gaining popularity in recent days. Last decades evidences many developments in edible coating. Chitosan based films are noteworthy to mention in this context. Chitin is obtained from natural sources, which is found to be present in the exo-skeleton of invertebrates. Molecular structure is composed of repeating units of 1, 4-linked 2-deoxy-2- acetamido-b-D-glucose. Chitosan is derived from Chitin by deacetylation process. Chitosan is found to exhibit moderate water vapour permeation characteristics and low oxygen permeability. However, the chitosan film possesses excellent mechanical strength like other commercially available common plastics (Kittur et al., 1998). Active packaging is very popular for the use in food packaging applications. It can be either flexible or rigid in nature. The flexibility depends on the type of polymer is in use for developing packaging materials.

Active packaging

Change in food handling and food distribution practices increase demand of improvement in shelf-life of the food items. In order to meet the demand active packaging has been endorsed as a new concept in the last two decades. Active packaging maintains the food condition by influencing physiological processes like respiration and chemical processes like lipid oxidation which are responsible for reducing shelf-life of the food items (Bodbodak and Rafiee 2016). Techniques for preservation and improving the shelf-life can be classified into three categories. These are scavenging, releasing and other systems. Scavenger systems are used majorly for absorbing oxygen, moisture, ethylene, taints and other undesirable compounds from head space of the packaged food items. The releasing systems are in use to add carbon dioxide, antioxidants, and various preservatives into the headspace of the package in order to extend the shelf-life of the packaged food items. Few examples of commercially available active packaging systems are marketed under the trade name of Ageless, Freshilizer, Freshmax, Freshpax Frsh Pack, Oxyguard, Zero$_2$, Bioka, Dri-Loc, Tenderpac, Biomaster, Agion, SANICO, Neupalon, Peakfresh, Evert-Fresh etc. (Biji et al., 2015) Base on the principal it can be classified into six categories: oxygen scavenger, moisture scavenger, ethylene scavenger, carbon dioxide absorbers, antimicrobial and antifungal.

Presence of oxygen in the package has many detrimental effects on the food items and causes "nutrient losses, color changes, off-flavour development, change in texture and microbial growth" (Bodbodak et al., 2016). Respiration rate as well as the ethylene production is also reported to be accelerated in presence of oxygen in the exposed atmosphere. Passive barrier packaging materials like high barrier packaging materials, multilayer laminates containing EVA (ethylene-co-vinyl alcohol) and aluminium foil, high barrier nanocomposites are few options adapted to control oxygen diffusion through packaging materials(Lagaron et al. 2004; Teixeira et al. 2011).Shin and others (2009) used an iron-based oxygen scravenger for extending the shelf-life of packaged meat items. The oxidative-induced color and flavor changes of the meatballs were found to be significantly low even after 9 months of storage at 30 °C (Shin et al., 2009). Matche and others (2011) used ascorbic acid with either iron or zinc to modify LLDPE films by blending procedure to facilitate oxidation process. Here, iron and zinc are used as transition metal catalyst where as ascorbic acid is used as reducing agent. Main function of ascorbic acid is to reactivate the oxidizing agent to prolong the duration of its action. Bread slices and bun wrapped in this film were reported to be sensory acceptable after 5-6 days (Matche and Sreekumar 2011). Beside this palladium are also used as oxygen scavengers. Here, Palladium was coated on PET/SiOx-films and used as oxygen scavenger. This film was evidenced to remove "2 vol % ofinitial oxygen in the headspace (160 cm^3) of MA-packed ham within 35 min after packaging". The unpackaged ham was evidenced to lose its redness in 2h. This characteristic allows the prevention of loss in redness of ham up to 21 days (Hutter et al., 2016). However, complete elimination of oxygen from the headspace in the package is impossible for any passive method. Oxygen scravengers are considered as most commercially viable solution for reducing oxygen concentration in the package. It maintains food quality by "decreasing food metabolism, reducing oxidative rancidity, inhibiting undesirable oxidation of labile pigments and vitamins, inhibiting enzymatic browning, controlling enzymatic discoloration, and inhibiting the growth of aerobic microorganisms" (Bodbodak et al., 2016).

Moisture also plays a key role in spoiling the food items. In order to protect the food items from moisture. Several moisture scavenger systems are in use which suppress moisture condensation inside the package. Such materials are evidenced to maintain the in-package RH below 97% (Mahajan PV, Rodrigues FA, Motel A, Leonhard A. 2008. Development of a moisture absorber for packaging of fresh mushrooms (Mahajan et al. 2008; Rux et al. 2015).

Carbon dioxide absorber and emitters are a special class of active packaging which are in use to suppress the microbial growth in the packaged food items like poultry, fish, meat, cheese and baked products (Biji et al., 2015). High CO_2 levels are recommended for these food items in order to extend the shelf-life. However, graded reduction in CO_2 due to solubilisation in food items creates a partial vacuum in the packages which in turn results in collapse of the package made by flexible films. Soluble gas stabilization is the process which is adopted to avoid such circumstances. Beside this, some manufacturers marketed CO_2 releaser (e.g. Ageless TM type G marketed by Mitsubishi Gas Chemical Co. Ltd.). CO_2 scavengers are used to remove excess CO_2 in hermetically sealed packs for fresh roasted coffees (Day 2008).

Ethylene functions as a natural plant growth hormone. It causes for fruit ripening and softening through accelerating the respiration process. Ethylene is evidenced to possess detrimental effect on shelf-life of fruits and vegetables. Thus, ethylene scavengers are used to reduce the concentration of ethylene in the food packages. Potassium permanganate ($KMnO_4$) immobolized on inert mineral is a popular choice as ethylene scravengers which are packed in sachets and used in the packages. Activated carbons impregnated with palladium catalyst also possess the ability to absorb ethylene gas. Few examples of this kind are Neupalon™,Sedo Mate, Hatofresh etc. "Electron deficient nitrogen containing trienes" encapsulated by "ethylenepermeable packaging" are also available for the use of scavenging ethylene from headspace of the packages (Biji et al., 2015). Some films like polystyrenes, polycarbonates, LDPE, PP are used to develop this type of packages. 1-methylcyclopropane is also used for reducing the effect of ethylene (Blankenship and Dole 2003).

Odour proof packages are also invented where flavours scavengers are used to prevent the cross-contamination of pungent odour during transportation. Some volatile compounds are also evidenced to accumulate in the packages due to degradation of certain food products. They are sulphides, amines and aldehydes. Depending on the type of volatile compounds materials are chosen to absorb the odour. Such as citric acid is used in sachets to absorb volatile amines produced due to breakdown of protein molecules present in fish muscle (Hoshino et al.1986). ANICO bags made of films containing ferrous salts and organic acid (e.g. ascorbic acid, citric acid etc.) is also found to be used as volatile amine scavenger (Rooney 1995).

Synthetic antioxidant such as butylatedhydroxyanisole was used as an ingredient to develop active packaging materials in order to protect meat against oxidation of lipid molecules. Beside this, use of plant extracts and essential oils are also used to eliminate many negative impacts on the food products. Many research articles were published on optimizing the concentration of antioxidant and its rate of release to enhance the extent of the protection effect of the packaging materials (Chen et al. 2012; Nerin et al. 2008).

Beside these, self-heating and self-venting packaging are also noteworthy to mention here. Self-heating packaging is used to heat food items without external heating sources. M. Hamasky filed a patent on self-heating container (Hamasaki and Chori 1988). V. Girish has also filed a patent on manufacturing of self-heating pouches in the year of 2018 (Girish 2018). Flameless Ration Heater is another system which contains amagnesium-iron compound. Saline solution is used to initiate an exothermic chemical reaction (Bartkowiak et al. 2015). Self-venting packages are used to maintain pressure in the pack. There are two types of self-vent systems used in packages: (i) valve system and (ii) non-valve systems. Valves are one way type which vent the steam. "Wipf" is an example of such valve. Nonvalve systems are of two types: (i) hole and (ii) Cohesive-seal or seal-break systems. The first one consists of micropores or larger holes or slits covered with a "barrierfilm, adhesive film tape, or adhesive label" (Schiffmann 2017). The second one contains "weak point at a tray/lid seal, or a material or device that is somehow microwave interactive and melts a hole or vent for steam release" (Schiffmann 2017). Microwave susceptor is used to evenly distribute the heating when the packages are subjected to microwave heating. Sira Crisp™, Smart Pouch are the examples of microwave subsectors which are commercially available (Biji et al., 2015). Ethanol emitters are also used to increase the preservation characteristics of the food packages. Ethicap™, Negamould™, Antimould102™, Oitech™ are the few examples of ethanol emitters (Rooney 1995).

Few examples of the additives used to develop active packaging solutions are discussed in Table 2 (yildirim et al., 2017).

Table 2: Few examples of additives used in active packaging materials and its roles, applications (Yildirim et al. 2017)

Additives	Function	Packaging applications	Food application
Iron	Oxidising agent	(i) Multilayer-container: PP/EVOH/OS/PP (ii) Mulitilayer film:PET/Alu/PE/OS/PE (iii) Laminates: PET/Alu/OS/PE ABSO2RB (iv) Sachets: Fe nanoparticles/Carbon/NaCl/$CaCl_2$	Meatballs, hot filled cheese spread, Roasted sunflower seeds, walnuts, bread
Zinc	Oxidising agent	Flexible film	Bread and Bun
Pd-deposited film		PET/SiOx/Pd	Cooked cured sliced ham
Ascorbic acid	Reducing agent	LLDPE-films	Bread and Bun
Eosin	Photosensitive dyes	Ethyl cellulose based cast film	Sunflower oil
Curcumin			Sunflower oil
Nupak		EVOH/ OS/EVA	Probiotic yogurt
Photosensitive dyes & reducible organic compounds		EVOH /OS/CPP laminate	Orange juice
Charcoal/ $PdCl_2$	Ethylene-absorbent	Paper packets in metal tray with glass cover	Fresh-cut kiwifruit, banana, spinach
Nano-Ag, Nano-TiO2 and kaolin		Polyethylene	Strawberry, Kiwifruit, Chinese bayberry, Chinese jujube
Nano-ZnO		As a coating on PVC film	Fresh-cut apple
Zeolite-based various alumino-silicate minerals		LDPE	Broccoli florets
Zeolite-based impregnated with KMnO4		HDPE	Kiwifruit
Zeolite fine particles of mordenite framework inverted-type zeolite		LDPE	Mango
Natural hydroscopic mineral		Commercial LDPE bags	Broccoli

Additives	Function	Packaging applications	Food application
Butylated hydroxyl toluene	Antioxidant	LDPE	Fresh sierra fish
α-Tocopherol		HDPE/EVOH/LDPE	Whole milk powder
Commercial mixtures of tocopherols		LDPE	Salmon
Quercetin, ascorbic acid, ferulic acid, and green tea extract		EVOH	Brined sardines
Thymol, carvacrol, and eugenol		Corn-zein-laminated linear LDPE film	Vacuum-packed beef patties
Resveratrol		Chitosan film	Pork sausages
Citric acid		Cornstarch/linear LDPE film	Vacuum- packed ground beef
Catechin, quercetin		EVOH	Fried peanuts, sunflower oil
Bentonite	Moisture scravenger	bags in the package	Mushroom
Sodium salt of polyacrylic acid		Powder in porous "tea bag" in sealed containers	Maize
Sodium chloride		(i) PE/foamed hygroscopic ionomer-NaCl/PE (ii) PP/foamed and stretched PP -NaCl/EVOH/PE	Mushrooms

Antimicrobial packaging also gains popularity as an active packaging solution which is used mainly to "reduce, inhibit or retard the growth of microorganisms that may be present in the packaged food or packaging material itself" (Biji et al., 2015). Antimicrobial coating is applied on the inner surface of the packaging materials. Natural antimicrobial agents e.g. extracts from spices like oregano, allispice, clove, cinnamon etc. are in use for developing antimicrobial packages. Beside this extracts of garlic, onion, radish mustard are also popular choice as antimicrobial additive. Antimicrobial additive can migrate to the surface of the packaging material or can be applied as direct coating on the surface of packaging materials. Major characteristic for the additives required to behave as antimicrobial materials is the ability to "extend the lag phase and reduce the growth phase of microorganisms" (Biji et al., 2015). In Japan, silver substituted zeolite is used as common antimicrobial agent as a thin layer which is laminated on the food contact surface. Zeomic®,Bactekiller are the examples of commercially available silver substituted zeolites. Antimicrobial enzymes (e.g. Lactoferrin, lactoperoxidase etc.), antimicrobialpeptides (e.g. cecropines, defensins, megainins), natural phenols (e.g. hydroquinons, catechins), metal (e.g. copper, silver) are also found to be incorporated into the polymer to develop antimicrobial characteristics (Hotchkiss 1997). There are few antimicrobial packaging systems that are developed by covalently immobilizing antimicronia additives on plastics to suppress microbial growth. A blend of polyethylene and polyamide (70:30) were evidenced to form a stablebond with nisin. It was observed to reduce the population of lactic acid bacteria significantly when in modified atmosphere packaging with 60% N2 and 10% CO2. This packaging system also can reduce level of *L. innocua* and *S. Aureus* (Kerry et al. 2006). A number of essential oils were recognized to have significant impact over anti-bacterial characteristics. Cinnamic acid, cinnamaldehyde, perillaldehyde, carvacrol, eugenol, thymol, cinnamon, marjoram are the few examples of such essential oils (Burt 2004; Fabian et al. 2006; Gutierrez et al. 2009).

Non-migratory bioactive polymer (NMBP) gains immense popularity as a class of polymers which possess "biological activity without the active components migrating from the polymer to the substrate (Steven and Hotchkiss 2003). This polymer finds applications for immobilised enzyme processing (Katchalski-Katzir 1993). Bioactive molecules elicit a response from living systems. The bioactive materials for which the response is desirable, goal is to inhibit microbial growth or flavour improvement. Enzymes, peptides, proteins and other organic compounds are the examples of bioactive compounds. Bioactive polymers can be formed by arresting bioactive molecules on synthetic polymers (Appendini and Hotchkiss 1997; Soares and Hotchkiss 1998). Non-migratory polymers are defined as the polymers from which no migration of the bioactive component occurs from the polymer into the surrounding. Such system is developed by covalent bond formation between polymer and the bioactive molecules or by entrapping the active components in the polymer matrix. NMBP has several advantages like (i) "improved stability of the bioactive substance, and concentration of the bioactive effect at a specific locus", (ii) as it has no migration possibilities, it has regulatory advantages, (iii) meeting the demand of safety minimally processed and preservative-free products and (iv) extending shelf-life of the packaged food items(Steven and Hotchkiss, 2003). Chitosan is most studied NMBP which was found to exhibit excellent anti-fungal property. It was also found to disrupt the outer shell of the bacteria (Tsai and Su 1999; Helander et al., 2001).

Eco-friendly packaging

In the current scenario of ever increasing plastic waste and its pollution, consumers are interested in earth friendly packaging options which possess ecological benefits along with the standard requirements for packaging of functional food and beverages (Hilton 2017). Bioplastics are developed by valorization of vegetable waste and can be blended with other synthetic polymers in order to improve their mechanical durability and gas barrier characteristics. A bioplastic made of polyvinyl alcohol and carrot was found to exhibit low permeability of oxygen (31.2 cm^3 μm m^{-2} day^{-1} kPa^{-1}) even from other synthetic polymers (Perotto et al. 2018). Liposomes gains popularity as carrier system to deliver various active compounds. For example, Nisin (An antimicrobial additive) were encapsulated in polyethylene oxide by electrospinning method. An edible film or coating is also noteworthy to mention here. Arcan and Yemenicioglu (2014) developed edible zein–fatty acid for controlled release abilities for many biologically active compounds. Such compounds are capable of showing antimicrobial and antioxidant effects (Brandelli et al., 2017). Biopolymer films made ofstarch, cellulose, chitosan and alginate are used to develop antimicrobial films. Silver nanoparticless and many natural products like bacteriocins, isothiocyanates, sorbic acid, and essential oils were used as antimicrobial additives for developing active packaging materials (Brandelli et al., 2017; Unalan et al., 2014). Recently, an investigation on barrier properties of eco-friendly polyester namely random copolymers of poly (propylene 1, 4-cyclohexanedicarboxylate) were reported Siracusa et al. (2018) (Siracusa et al. 2018).This type of aliphatic polyester is biodegradable and bio-based. It is found to offer several advantages like "excellent tensile strength, stiffness and impact properties, high thermal stability, and easily attackable by microorganisms" (Siracusa et al., 2018). Gas transmission rate of oxygen and carbon dioxide for this film was found to be in line with the PET, PP and HDPE. In a very recent study, Zhang and his co-workers developed bioplastic using hazardous high-salt microalgal (*Spirulina* sp.) residues (Zhang et al. 2019). Polyvinyl alcohol of molecular weight 3100 kDa is used as a base polymer materials which is known as biodegradable as well as hydrophillic polymer (Galiano et al. 2018). Here the biomass is used as filler only. The developed biofilm is found to exhibit a decent mechanical durability and good water resistance which is prime requirement for packaging applications. Polysaccharide electrolyte complexes as eco-friendly coating materials for packaging application, is also noteworthy to mention here. Chitosan, carboxymethyl cellulose, starch, pectin, sea wood extracts are used to synthesize the coating material which can be used for food packaging applications (Umaraw and Verma 2015). Alginates isolated from the cell walls of brown algae (e.g. *Laminaria digtata, Ascophyllum nodosum*) are used to develop edible film for meat packaging applications. It is also used in pharmaceuticals and medicine for encapsulation or as part of prolonged-release drugs. In addition, it also finds its use in bandages due to its special characteristics like bacteriostatic, fungistatic, antiviral, non-toxic, intensely absorbent, non-allergic, hemostatic and biocompatible Gheorghita (Puscaselu et al. 2020). According to Moustafa et al. (2018), the attention is drifted towards green packaging based on biodegradable composites. 100% biodegradability and ability to complete breakdown to carbon dioxide, water and humus are the two major features that makes it popular as eco-friendly material (Moustafa et al. 2019).Polyvinyl alcohol/rice starch/silk fibroin-blended films were found to exhibits agreeable results in terms of oxygen permeability as well as mechanical properties (Kuchaiyaphum et al. 2013). Xylan, a form of hemicellulose naturally occurring in plant cell and some algae are used to develop xylophane film which is also very popular as packaging materials. In addition, many plant and animal derived proteins also find its use in developing eco-friendly packaging materials. Glutan, Soy protein isolates, Whey and Zein are the examples of plant derived proteins which find promising applications in packaging. Animal derived proteins like Casein, Keratin, Collagen, Gelatin also gain popularity in recent days for many packaging applications. Paper and paper board being biodegradable, biocompatible can be an excellent choice for the use as eco-friendly packaging materials but suffers from serious drawbacks like permeability to moisture and gas, easy to tear, inferior wet strength (Singh et al. 2011). Starch is abundantly available in nature and composed of mainly two polymers amylose and amylopectin. Semicrystalline starch granules of diameters 1-150 μm and a crystalline index of 15-45% are a rich choice for producing edible films which can provide adequate protection to food items (Pelissari et al. 2019). Starch based biocomposites are also noteworthy to mention here which provides improved shelf-life for the food items. Various fibrous fillers like natural fibers and nonfibrous fillers like calcium carbonate, silver nanoparticles, clays etc.to improve its physical properties as well as to incorporate new functionalities. Starch based biocomposites or nanocomposites usually obtained by "solution casting" method after dispersing the filler in s suitable solvent. Glycerol is used mainly as plasticizer to obtain TPS with incorporated nanosilver, titanium dioxide, nanoclays, nano-zinc oxide, nano-magnesium hydroxide, silicon dioxide as filler components(Abreu et al. 2015; Goudarzi et al. 2017; Hassani et al. 2014; Perotti et al. 2017; Pelissari et al. 2019.

Conclusion

Quality of naturally occurring functional foods and nutraceuticals, is observed to be deteriorated with time. Thus protecting such food items pose a great challenge to the industry and can be seen as main obstacle in marketing of this type of products. Packaging evolves as a solution to this problem which is not only found to offer protection and prevention to the packaged food items but also improves presentation and enriches the aesthetic appearance of the product. This chapter briefs about the various packaging technology and related latest innovations which are useful for packing functional food items and nutraceuticals. As concern towards protecting environment and reducing the waste generation increases, the consumer demand is drifted towards eco-friendly packaging solutions. Hence, few advancement and important eco-friendly packaging solutions that can be used for packaging of functional food items are also briefed in this chapter.

Acknowledgement: Authors acknowledge Indian Institute of Packaging, Mumbai for providing support in writing the book chapter.

References

Abreu, A.S., Oliveira, M., Sá, A., Rodrigues, R.M., Cerqueira, M.A., Vicente, A.A., Machado, A.V., 2015. Antimicrobial nanostructured starch based films for packaging. Carbohydrate Polymers 129, 127e134.'

Appendini, P., Hotchkiss, J.H. (1997) Immobilisation of lysozyme on food contact polymers as potential anti-microbial films. Packag. Technol. Sci. 10, 271–9;

Bandyopadhyay B, Mandal NC (2014) Probiotics, prebiotics and synbiotics-in health improvement by modulating gut microbiota: the concept revisited. Int J Curr Microbiol Appl Sci 3:410–420.

Bartkowiak, A., Mizielińska, M., Sumińska, P., Romanowska-Osuch, A., & Lisiecki, S. (2015). Innovations in Food Packaging Materials. Emerging and Traditional Technologies for Safe, Healthy and Quality Food, 383–412. doi:10.1007/978-3-319-24040-4_19

Bell, L.N. 2006. Nutraceutical stability concerns and shelf life testing. In Handbook of Nutraceuticals and Functional Foods, ed. R.E.C. Wildman. CRC Press, Boca Raton, FL.

Beyers, M., & Thomas, A. C. (1979). Γ Irradiation of subtropical fruits. 4. Changes in certain nutrients present in mangoes, papayas, and litchis during canning, freezing, and Γ irradiation. Journal of Agricultural and Food Chemistry, 27, 48–51.

Biji, K.B., Ravishankar, C.N., Mohan, C.O. and Gopal, T.S., 2015. Smart packaging systems for food applications: a review. Journal of food science and technology, 52(10), pp.6125-6135.

Binninger, A.S., 2017. Perception of naturalness of food packaging and its role in consumer product evaluation. Journal of Food Products Marketing, 23(3), pp.251-266.

Blankenship SM, Dole JM (2003) 1-Methylcyclopropene: a review. Postharvest Biol Tec 28:1–25.

Bodbodak, S., & Rafiee, Z. (2016). Recent trends in active packaging in fruits and vegetables. Eco-Friendly Technology for Postharvest Produce Quality, 77–125.

Brandelli, A., Brum, L. F. W., & dos Santos, J. H. Z. (2017). Nanostructured bioactive compounds for ecological food packaging. Environmental Chemistry Letters, 15(2), 193–204.

Burt S (2004) Essential oils: their antibacterial properties and potential applications in foods—a review. Int J Food Microbiol 94:223–253.;

Chen, X., Lee, D. S., Zhu, X., & Yam, K. L. (2012). Release kinetics of tocopherol and quercetin from binary antioxidant controlled-release packaging films. Journal of Agricultural and Food Chemistry, 60, 3492-3497.

Day BPF (2008) Active packaging of food. In: Kerry J, Butler PJ (eds) Smart packaging technologies for fast moving consumer goods. Wiley and sons Ltd, West Sussex, England.

Deschênes, L., 2015. Packaging Functional Foods: From Basic Requirements to Nano Perspectives. Functional Food Ingredients and Nutraceuticals: Processing Technologies, 13, p.409.

Fabian D, Sabol M, Domaracká K, Bujňáková D (2006) Essential oils—their antimicrobial activity against Escherichia coli and effect on intestinal cell viability. Toxicol in Vitro 20:1435–1445;

Galiano, F., Briceno, K., Marino, T., Molino, A., Christensen, K.V., Figoli, A., 2018. Advances in biopolymer-based membrane preparation and applications. J Membrane Sci., 564, 562-586.

Gheorghita (Puscaselu), Gutt, & Amariei. (2020). The Use of Edible Films Based on Sodium Alginate in Meat Product Packaging: An Eco-Friendly Alternative to Conventional Plastic Materials. Coatings, 10(2), 166.

Gil, E. C., A. I. Colarte, J. L. L. Sampedro, and B. Bataille. 2008. Subcoating with Kollidon VA 64 as Water Barrier in a New Combined Native Dextran/HPMC-Cetyl Alcohol Controlled Release Tablet. European Journal of Pharmaceutics and Biopharmaceutics 69: 303–11.

Girish, V., Extovate Venture LLP, 2018. Self-heating pouch and method of manufacture thereof. U.S. Patent Application 15/757,366.

González-Herrera, S. M., Bermúdez-Quiñones, G., Ochoa-Martínez, L. A., Rutiaga-Quiñones, O. M., & Gallegos-Infante, J. A. (2020). Synbiotics: a technological approach in food applications. Journal of Food Science and Technology.doi:10.1007/s13197-020-04532-0.

Goudarzi, V., Ghahfarrokhi, I.S., Ghazvini, A.B., 2017. Preparation of ecofriendly UV-protective food packaging material by starch/TiO2 bionanocomposite: characterization. International Journal of Biological Macromolecules 95, 306e313.

Gupta, R. C., Srivastava, A., Lall, R. (Eds.). (2019). Nutraceuticals in Veterinary Medicine. doi:10.1007/978-3-030-04624-8.

Gutierrez J, Barry-Ryan C, Bourke P (2009) Antimicrobial activity of plant essential oils using food model media: efficacy, synergistic potential and interactions with food components. Food Microbiol 26:142–150.

Hamasaki, M., Chori Co Ltd, Hama Corp Co Ltd and FUKUBI KAGAKU KOGYO KK, 1988. Self-heating container. U.S. Patent 4,762,113.

Hassani, F.S., Nafchi, A.M., 2014. Preparation and characterization of bionanocomposite films based on potato starch/halloysite nanoclay. International Journal of Biological Macromolecules 67, 458e462.

Helander, I.M., Nurmiaho-Lassila, E.L., Ahvenainen, R., Rhoades, J., Roller, S. (2001) Chitosan disrupts the barrier properties of the outer membrane of Gram-negative bacteria. Int. J. Food Microbio. 71, 235–44.

Hilton, J. (2017). Growth patterns and emerging opportunities in nutraceutical and functional food categories: market overview. Developing New Functional Food and Nutraceutical Products, 1–28.

Hoshino A, Osanai T, inventors (1986) Packaging films for deodorization. Japanese patent 86209612.

Hotchkiss J (1997) Food packaging interactions influencing quality and safety. Food Addt Contam 14(6):601–607.; Appendini P, Hotchkiss JH (2002) Review of antimicrobial food packaging. Innov Food Sci Emerg 3:113–126.

Hutter S, R¨uegg N, Yildirim S. 2016. Use of palladium based oxygen scavenger to prevent discoloration of ham. Food Pack Shelf Life 8:56–62.

Kamala, K. and Kumar, V.P., 2018. Food Products and Food Contamination. In Microbial Contamination and Food Degradation (pp. 1-19). Academic Press.

Katchalski-Katzir, E. (1993) Immobilized enzymes – learning from past successes and failures. TIBTECH 11, 471–8.; MOSBACH, K. (1980) Immobilized enzymes. Trends Biochem. Sci. 5, 1–3.

Kerry JP, O'Grady MN, Hogan SA (2006) Past, current and potential utilisation of active and intelligent packaging systems for meat and muscle-based products: a review. Meat Sci 74:113–130

Kittur, F.S., Kumar, K.R. and Tharanathan, R.N., 1998. Functional packaging properties of chitosan films. Zeitschrift für Lebensmitteluntersuchung und-Forschung A, 206(1), pp.44-47.

Korbekandi, H. Mortazavian, A.M. Iravani, S. Technology and stability of probiotic in fermented milks. In N. Shah, A. G. Cruz, and J. A. F. Faria (Eds.), Probiotic and prebiotic foods: Technology, stability and benefits to the human health (2011) pp. 131-169.

Kuchaiyaphum, P., Punyodom, W., Watanesk, S. and Watanesk, R. (2013). Composition optimization of polyvinyl alcohol/rice starch/silk fibroin-blended films for improving its eco-friendly packaging properties. Journal of Applied Polymer Science, 129(5), 2614–2620.

Lagaron, J.M., Catala, R., Gavara, R., 2004. Structural characteristics defining high barrier properties in polymeric materials. Mater. Sci. Technol. 20, 1–7.;

Mahajan, P. V., Rodrigues, F. A., Motel, A., and Leonhard, A. (2008). Development of a moisture absorber for packaging of fresh mushrooms (Agaricus bisporous). Postharvest Biology and Technology, 48(3), 408-414.

Mahanjan, R., A. Templeton, A. Harman, R. Reed, and R. T. Chern. 2005. The Effect of Inert Atmospheric Packaging on Oxidative Degradation in Formulated Granules. Pharmaceutical Research 22: 128–40.

Matche RS, Sreekumar RK, Raj B. 2011. Modification of linear low-density polyethylene film using oxygen scavengers for its application in storage of bun and bread. J Appl Polym Sci 122(1):55–63.

Mizielińska, M., Sobecka, K., Jarosz, M., Urbański, D., Stobińska, M., Łukawska, B., Olchawa, E. and Bartkowiak, A., 2017. The influence of immobilization and packaging on the viability of probiotics stored at 25° C. World Scientific News, 77(2), pp.124-143.

Moustafa, H., Youssef, A. M., Darwish, N. A., & Abou-Kandil, A. I. (2019). Eco-friendly polymer composites for green packaging: Future vision and challenges. Composites Part B: Engineering.

Nerin, C., Tovar, L., and Salafranca, J. (2008). Behaviour of a new antioxidant active film versus oxidizable model compounds. Journal of Food Engineering, 84, 313-320.

Oyetade OA, Oyeleke GO, Adegoke BM et al (2012) Stability studies on ascorbic acid (vitamin C) from different sources. IOSR J Appl Chem 2(4):20–24

Pacheco-Palencia, L. A., Duncan, C. E. and Talcott, S. T. (2009). Phytochemical composition and thermal stability of two commercial açai species, Euterpeoleracea and Euterpeprecatoria. Food Chemistry, 115, 1199–1205.

Pamplona-Roger, G. D. 1998. Encyclopedia of Medicinal Plants. Edited by G. X. Francesc. Madrid: Editorial Safeliz, S.L., vol. 1, 27–51.

Pascall, M.A., 2011. 13 Chapter the packaging of Nutraceuticals Derived from plants. Handbook of Nutraceuticals Volume II: Scale-Up, Processing and Automation, 2, p.313.

Pelissari, F. M., Ferreira, D. C., Louzada, L. B., dos Santos, F., Corrêa, A. C., Moreira, F. K. V., and Mattoso, L. H. (2019). Starch-Based Edible Films and Coatings. Starches for Food Application, 359–420.

Perotti, G.F., Kijchavengkul, T., Auras, R.A., Constantino, V.R.L., 2017. Nanocomposites based on cassava starch and chitosan-modified clay: physico-mechanical properties and biodegradability in simulated compost soil. Journal of the Brazilian Chemical Society 28, 649e658.

Perotto, G., Ceseracciu, L., Simonutti, R., Paul, U.C., Guzman-Puyol, S., Tran, T.N., Bayer, I.S. and Athanassiou, A., 2018. Bioplastics from vegetable waste via an eco-friendly water-based process. Green Chemistry, 20(4), pp.894-902.

Prakash, S., Jha, S. K., & Data, N. (2004). Performance evaluation of blanched carrots dried by three different driers. Journal of Food Engineering, 62, 305–313.

Rawson, A., Patras, A., Tiwari, B.K., Noci, F., Koutchma, T., Brunton, N. 2011. Effect of thermal and nonthermal processing technologies on the bioactive content of exotic fruits and their products: Review of recent advances. Food Research International, 44(7):1875–1887.

Raybaudi-Massilia, R. M., Mosqueda-Melgar, J., and Tapia, M. S. (2010). Edible coatings as carriers of food additives on fresh-cut fruits and vegetables. Stewart Postharvest Review, 3(3), 1-

Rooney ML (1995) Overview of active food packaging. In: Rooney ML (ed) Active food packaging. Blackie Academic & Professional, Glasgow, pp 1–37.

Rooney ML (ed) (1995) Active food packaging. Chapman & Hall, London.

Rux G, Mahajan PV, Geyer M, Linke M, Pant A, Saengerlaub S, Caleb OJ. 2015. Application of humidity-regulating tray for packaging of mushrooms. Postharvest Biol Tec 108:102–10.

Saint-Eve, A., Levy, C., Le Moigne, M., Ducruet, V., Souchon, I., 2008. Quality changes in yogurt during storage in different packaging materials. Food Chem. 110 (2), 285–293)

Schiffmann, R. (2017). Packaging for microwave foods. The Microwave Processing of Foods, 273–299.

Senaka Ranadheera, C., Evans, C. A., Adams, M. C., and Baines, S. K. (2013). Production of probiotic ice cream from goat's milk and effect of packaging materials on product quality. Small Ruminant Research, 112(1-3), 174–180.

Shahidi, F., Naczk, M. 2004. Phenolics in Food and Nutraceuticals. Taylor & Francis Group, Boca Raton, FL.

Shin Y, Shin J, Lee Y. 2009. Effects of oxygen scavenging package on the quality changes of processed meatball product. Food Sci Biotechnol 18(1):73–8.

Singh, A., Sharma, P. K., and Malviya, R. (2011). Eco friendly pharmaceutical packaging material. World Applied Sciences Journal, 14(11), 1703-1716.

Siracusa, V., Genovese, L., Ingrao, C., Munari, A., and Lotti, N. (2018). Barrier Properties of Poly (Propylene Cyclohexanedicarboxylate) Random Eco-Friendly Copolyesters. Polymers, 10(5), 502.

SOARES, N.D.F.F. and HOTCHKISS, J.H. (1998) Naringinase immobilization in packaging films for reducing naringin concentration in grapefruit juice. J. Food Sci. 63, 61–5.

Steven, M. D., and Hotchkiss, J. H. (2003). 5-Non-migratory bioactive polymers (NMBP) in food packaging A2-Ahvenainen, Raija. Novel Food Packaging Techniques: Woodhead Publishing, 71-102.

Teixeira, V., Carneiro, J., Carvalho, P., Silva, E., Azevedo, S., Batista, C., 2011. High barrier plastics using nanoscale inorganic films. In: Lagarón, J.M. (Ed.), Multifunctional and Nanoreinforced Polymers for Food Packaging. Woodhead Publishing, Cambridge, UK, pp. 285–315.

Tripathi, M.K. and Giri, S.K. Probiotic functional foods: Survival of probiotics during processing and storage. J. Funct. Foods.9 (2014) 225-241;

TSAI, G.J. and SU, W.H. (1999) Antibacterial activity of shrimp chitosan against Escherichia coli. J. Food Prot. 62, 239–43;

Umaraw, P., and Verma, A. K. (2015). Comprehensive review on application of edible film on meat and meat products: An eco-friendly approach. Critical Reviews in Food Science and Nutrition, 57(6), 1270–1279.

Unalan UI, Cerri G, Marcuzzo E, Cozzolino CA, Farris S (2014) Nanocomposite films and coatings using inorganic nanobuilding blocks (NBB): current applications and future opportunities in the food packaging sector. RSC Adv 4:29393–29428.

Venir, E., Maltini, E. 2013. Relevance of physical properties in the stability of plantbased food products. Indian Journal of Experimental Biology, 51(11):894–904.

Yildirim, S., Röcker, B., Pettersen, M. K., Nilsen-Nygaard, J., Ayhan, Z., Rutkaite, R. Radusin, T., Suminska, P., Marcos, B. & Coma, V. (2018). Active packaging applications for food. Comprehensive Reviews in Food Science and Food Safety, 17(1), 165-199.

Zhang, C., Wang, C., Cao, G., Wang, D., & Ho, S.-H. (2019). A sustainable solution to plastics pollution: an eco-friendly bioplastic film production from high-salt contained Spirulina sp. residues. Journal of Hazardous Materials, 121773.